PHARMACOLOGICAL APPROACHES IN CANCER TREATMENT

PHARMACOLOGICAL APPROACHES IN CANCER TREATMENT

A COMPREHENSIVE GUIDE

Edited by

Sadhna Aggarwal

Department of Radiation Oncology, University of Texas MD Anderson Cancer Center, TX, United States

Phulen Sarma

Department of Pharmacology, AIIMS, Guwahati, Assam, India

ELSEVIER

ACADEMIC PRESS

An imprint of Elsevier

Academic Press is an imprint of Elsevier
125 London Wall, London EC2Y 5AS, United Kingdom
50 Hampshire Street, 5th Floor, Cambridge, MA 02139, United States

ISBN: 978-0-443-38307-6

For information on all Academic Press publications
visit our website at https://www.elsevier.com/books-and-journals

Publisher: Megan Ball
Acquisitions Editor: Linda Buschman
Editorial Project Manager: Sumit Yadav
Production Project Manager: Omer Mukthar
Cover Designer: Mark Rogers

Typeset by Integra Software Services

Contents

7. Palliative care in oncology: Symptom management and supportive therapies

Gautam Sarma, Jyotiman Nath, and Partha Pratim Medhi

8. Adverse effects of cancer pharmacotherapy: Prevention and management

Rituparna Chetia and Manas DubeyInfections in cancer

9. Preclinical studies: Translational research in drug development

Sankar Jyoti Bora, Santeena Chensula, Anusuya Bhattacharyya, Phulen Sarma, and Parag Deb RoyIntroduction

Contributors

Sadhna Aggarwal Department of Radiation Oncology, University of Texas MD Anderson Cancer Center, Houston, TX, United States

Sowvik Bag M.M. College of Pharmacy, Maharishi Markandeshwar (Deemed to be University), Ambala, Haryana, India

Nitin Bansal Department of Pharmaceutical Sciences, Chaudhary Bansi Lal University, Bhiwani, India

Seema Bansal M.M. College of Pharmacy, Maharishi Markandeshwar (Deemed to be University), Ambala, Haryana; Department of Pharmacology, Post Graduate Institute of Medical Education and Research, Chandigarh, India

Anusuya Bhattacharyya Department of Ophthalmology, AIIMS, Guwahati, India

Sankar Jyoti Bora Department of Pharmacology, AIIMS, Guwahati, India

Rishabh Chaudhary M.M. College of Pharmacy, Maharishi Markandeshwar (Deemed to be University), Ambala, Haryana, India

Madhavrao Chavan Department of Pharmacology, Government Medical College, Chandrapur, Maharashtra; Department of Pharmacology, AIIMS, Mangalagiri, Andhra Pradesh, India

Santenna Chenchula Department of Pharmacology, AIIMS, Bhopal, Madhya Pradesh, India

Santeena Chensula Department of Clinical Pharmacology, AIIMS, Bhopal, India

Rituparna Chetia Department of Medical Oncology/Hematology, Medicine, Clinical Hematology, AIIMS, Guwahati, India

Manas Dubey Medical Oncology, Tata Memorial Centre, Guwahati; Consultant Medical and Hemato-Oncologist, Kamla Nehru Memorial Hospital, Prayagraj, India

Jitender Gairola Consultant-Centre of Excellence for Use of Drones in Medicine, AIIMS, Rishikesh, India

Bipasha Kalita Department of Pharmacology, AIIMS, Guwahati, India

Sadiyaakhtar S. Kazi Department of Pharmacology, Government Medical College, Chandrapur, Maharashtra, India

Harish Kumar Department of Pharmacology, AIIMS GUWAHATI, India

Department of Pharmacology, AIIMS-GUWAHATI, New Delhi, India

Anitha Kuttiappan Department of Pharmacology, School of Pharmacy and Technology Management (SPTM), SVKM's NMIMS Deemed-to-University, Shirpur, India

Partha Pratim Medhi Department of Radiation Oncology, AIIMS, Guwahati, Assam, India

Jyotiman Nath Department of Radiation Oncology, Dr. B. Borooah Cancer Institute, Guwahati, Assam, India

Parag Deb Roy Integrated Disease Surveillance Programme, Office of the Joint Director of Health Services, Guwahati, India

Gautam Sarma Department of Radiation Oncology, AIIMS, Guwahati, Assam, India

Indrani Devi Sarma Department of Pharmacology, AIIMS, Guwahati, India

Phulen Sarma Department of Pharmacology, AIIMS, Guwahati; Department of Pharmacology, Post Graduate Institute of

Medical Education and Research, Chandigarh, India

Phulen Sharma Department of Pharmacology, AIIMS, Guwahati, Assam, India

Arumilli Swetha Department of Pharmaceutics, Aditya College of Pharmacy, Surampalem, Andhra Pradesh, India

CHAPTER

1

Targeted therapies: Mechanisms and applications

Harish Kumar[a], *Anusuya Bhattacharyya*[b], *and Phulen Sarma*[c]

[a]Department of Pharmacology, AIIMS GUWAHATI, India

[b]Department of Ophthalmology, AIIMS, Guwahati, India

[c]Department of Pharmacology, AIIMS, Guwahati, India

Introduction

Targeted therapy is a revolutionary approach in cancer therapy that is designed to focus on specific molecular targets that are essential for tumor growth and survival. In contrast to traditional chemotherapy, which broadly targets all rapidly dividing cells, targeted therapies focus on disrupting cancer growth by specifically interfering with certain proteins, signaling pathways, or cellular mechanisms.

Targeted therapy, also known as molecular targeted therapy, is a form of treatment designed to inhibit the growth of cancer cells by disrupting specific molecular mechanisms essential for carcinogenesis and tumor development. Unlike traditional chemotherapy, which broadly targets all rapidly dividing cells, this approach focuses on precise molecular targets within cancer cells (Joo et al., 2013).

These include blocking growth signals [e.g., epidermal growth factor receptor (EGFR) inhibitors], preventing angiogenesis [e.g., vascular endothelial growth factor (VEGF) inhibitors], inducing apoptosis, or enhancing immune responses (e.g., immune checkpoint inhibitors) (Hanahan and Weinberg, 2011; Collins and Workman, 2006).

Targeted therapy marks a transformative approach in cancer treatment by concentrating on the molecular and cellular alterations unique to cancer. In contrast to traditional chemotherapy, which broadly targets all rapidly dividing cells, targeted therapies are designed to:

1. Interrupt specific pathways critical for cancer cell growth and survival.
2. Exploit unique molecular markers of cancer cells, minimizing harm to normal cells (Joo et al., 2013).

Pharmacological Approaches in Cancer Treatment
https://doi.org/10.1016/B978-0-443-38307-6.00011-9

These therapies are developed based on a comprehensive understanding of cancer genomics and proteomics, enabling personalized approaches to treatment.

Targeted therapy represents a transformative approach in cancer treatment, offering precision by disrupting specific molecular pathways essential for tumor growth. This strategy, under the umbrella of personalized medicine, has demonstrated notable success with therapies such as bevacizumab and olaparib for ovarian cancer, as well as trastuzumab combinations and mTOR (mechanistic target of rapamycin) inhibitors for breast cancer (Joo et al., 2013).

Key features of targeted therapy

1. Precision: Targets specific molecules associated with cancer, sparing most normal cells (Malone et al., 2020).
2. Molecular basis: Focuses on genetic, protein, or signaling alterations in cancer cells (Markowitz and Bertagnolli, 2009).
3. Less toxic: Generally associated with fewer side effects compared to traditional chemotherapy (Cancer Treatment Centers of America, 2024).
4. Personalization: Biomarker testing (e.g., genetic profiling) is often used to identify which patients will benefit from targeted treatments, e.g., breast cancer treatment (Committee on Policy Issues in the Clinical Development and Use of Biomarkers for Molecularly Targeted Therapies, n.d.; 2024).

Types of targeted therapy

Monoclonal antibodies

Monoclonal antibodies are specifically designed to target and bind to unique antigens present on the surface of cancer cells. They can inhibit signals that stimulate cancer cell growth or tag the cells for elimination by the immune system.

Example: Trastuzumab (Herceptin) specifically targets breast cancer cells that overexpress the HER2 protein. It has been demonstrated to lower the likelihood of recurrence and reduce mortality rates associated with HER2-positive breast cancer (Bradley et al., 2021).

Tyrosine kinase inhibitors

Tyrosine kinase inhibitors (TKIs) block the action of tyrosine kinases, enzymes that promote cell division. By inhibiting these enzymes, the proliferation and metastasis of cancer cells can be delayed or halted.

Example: Imatinib is used for chronic myeloid leukemia (CML) (Druker et al., 2006; Druker et al., 2001).

Angiogenesis inhibitors

Antiangiogenic drugs inhibit the development of new blood vessel formation, a process that tumors rely on for growth and metastasis.

Example: Bevacizumab is a drug that inhibits angiogenesis and is used in various cancers, such as colorectal and lung cancer (Ferrara et al., 2004; Díaz-Rubio and Schmoll, 2005).

Hormone therapies

These therapies target cancers that are driven by hormones, such as breast or prostate cancers. They either block the hormones or lower their levels in the body (Chaput and Sumar, 2022).

Example: Tamoxifen is used for estrogen receptor-positive breast cancer (Fisher et al., 1998).

Immune checkpoint inhibitors

Although part of immunotherapy, immune checkpoint inhibitors can be considered a form of targeted therapy, as they target the proteins used by cancer cells to evade immune system detection (Shiravand et al., 2022; http://National Cancer Institute, n.d.).

Example: Pembrolizumab targets the PD-1/PD-L1 pathway (Kwok et al., 2016).

Poly (ADP-ribose) polymerase inhibitors

These inhibitors are designed to target cancer cells that depend on the enzyme poly (ADP-ribose) polymerase (PARP) for DNA repair, especially in cancers with mutations in the BRCA1 or BRCA2 genes (Rose et al., 2020).

Example: Olaparib is used for BRCA (BReast CAncer genes)-mutated breast and ovarian cancers (Fong et al., 2009).

Mechanism of action of targeted therapy in cancer

The mechanisms of action of targeted therapies can be broken down into several key processes, each designed to interfere with specific pathways or cellular functions that support cancer growth and spread. Some mechanisms of action of targeted cancer therapy are given in the following.

Blocking cancer cell growth signals

Cancer cells often have mutations or overexpressed proteins on their surface that lead to uncontrolled growth signals. Under normal conditions, growth signals are tightly regulated and only initiate cell division when needed. However, in cancer cells, these signals may be constant, causing cells to divide uncontrollably. Targeted therapies can interfere with these signals by inhibiting the receptors or proteins that drive this abnormal cell division (Sever and Brugge, 2015).

Mechanism: Targeted therapies block or inhibit the action of specific receptors or proteins that promote cell growth. EGFR is often found to be overexpressed or mutated in various cancers, including nonsmall cell lung cancer (NSCLC). Inhibiting EGFR signaling can reduce the growth and spread of these cancers (Bethune et al., 2010).

Example: Erlotinib is a TKI that targets the EGFR receptor, blocking downstream signals that promote cell proliferation in NSCLC and pancreatic cancer (Bethune et al., 2010; Starling et al., 2006).

Inducing cancer cell death (apoptosis)

Cancer cells, unlike normal cells, are frequently able to bypass programmed cell death (apoptosis), which enables them to persist and grow indefinitely. Some targeted therapies are designed to induce apoptosis in cancer cells by restoring the mechanisms that regulate cell death (Fernald and Kurokawa, 2013).

Mechanism: Targeted therapies can promote apoptosis by blocking antiapoptotic proteins or activating proapoptotic pathways in cancer cells (Pfeffer and Singh, 2018). For example, proteasome inhibitors disrupt the balance of proteins in cancer cells, leading to the accumulation of defective proteins that trigger cell death (Nunes and Annunziata, 2017).

Example: Bortezomib is a proteasome inhibitor used in multiple myeloma that induces apoptosis by inhibiting the proteasome's function (Nunes and Annunziata, 2017; San Miguel et al., 2008).

Blocking angiogenesis (blood vessel formation)

To grow beyond a certain size, tumors require an adequate blood supply to deliver oxygen and nutrients. Angiogenesis refers to the process by which tumors promote the formation of new blood vessels. Targeted therapies can inhibit angiogenesis, effectively "starving" the tumor and limiting its growth (http://:~:text=wet%20macular%20degeneration,2024).

Mechanism: Angiogenesis inhibitors work by blocking the function of proteins, such as VEGF, which play a crucial role in the process of new blood vessel formation. By blocking these signals, targeted therapies prevent the tumor from forming the necessary blood supply.

Example: Bevacizumab (Avastin) is an angiogenesis inhibitor that works by binding to VEGF A, preventing its interaction with VEGF receptors on endothelial cells. This blockage disrupts the formation of new blood vessels within tumors, which slows down tumor growth and metastasis. It is commonly used to treat cancers such as colorectal and lung cancer (http://:~:text=wet%20macular%20degeneration, 2024; Hurwitz et al., 2004; Gerriets and Bevacizumab, 2023).

Delivering cytotoxic agents to cancer cells

Certain targeted therapies are developed to direct cytotoxic agents specifically to cancer cells. This strategy reduces harm to adjacent healthy tissues by ensuring that toxic substances are selectively delivered to tumor cells that display particular markers or receptors (Li et al., 2023).

Mechanism: Monoclonal antibodies can be conjugated to cytotoxic drugs, such as chemotherapy agents or radioactive substances. These antibody–drug conjugates (ADCs) specifically bind to antigens expressed on the surface of cancer cells, allowing the toxic agent to enter the cell and kill it (Lo Nigro et al., 2019).

Example: Gemtuzumab ozogamicin is an ADC that targets CD33 on leukemia cells and delivers a cytotoxic chemotherapy agent directly to the cancer cell, killing it while sparing normal cells (Clinical Trials Using Gemtuzumab Ozogamicin, n.d.; Rowe and Löwenberg, 2013).

Inhibiting hormone signaling

Certain cancers, such as breast and prostate cancers, rely on hormones (e.g., estrogen, testosterone) for growth and survival. Targeted therapies in the form of hormone therapies can block hormone production or action on cancer cells, thereby slowing or stopping tumor growth (Ulm et al., 2019).

Mechanism: Hormone therapies can either block hormone receptors or inhibit the production of specific hormones. For example, antiestrogens interact with estrogen receptors on breast cancer cells, inhibiting the binding of estrogen and thereby preventing its ability to promote the growth of cancer cells (Ulm et al., 2019).

Example: Tamoxifen is an antiestrogen used in estrogen receptor-positive breast cancer to block estrogen from binding to cancer cell receptors (Fisher et al., 2005).

Targeting cancer-specific mutations

Many cancers harbor specific mutations that drive tumorigenesis. Targeted therapies can selectively inhibit the mutated proteins or pathways associated with these mutations, offering a highly specific treatment approach (Min and Lee, 2022).

Mechanism: Targeted therapies can block the function of mutated oncogenes or tumor suppressors that are responsible for the growth of cancer cells. This approach is particularly useful when the mutations are well characterized, as in the case of BRAF mutations in melanoma.

Example: Vemurafenib is a targeted treatment that blocks the BRAF protein, particularly the BRAF V600E mutation, which is frequently present in melanoma. By blocking this mutated BRAF protein, vemurafenib disrupts the signaling pathway that promotes cancer cell proliferation, resulting in tumor shrinkage and clinical improvement (Chapman et al., 2011).

Application targeted therapy in cancer

Some of the applications of targeted therapies in clinical use are given in the following.

Breast cancer

Hormone receptor-positive breast cancer

Drugs such as tamoxifen (Farrar, Jacobs and Tamoxifen, 2023) and aromatase inhibitors (e.g., anastrozole) block estrogen pathways in estrogen receptor-positive cancers (Cuzick et al., 2010).

HER2-positive breast cancer

Therapies such as trastuzumab and pertuzumab target HER2 protein, reducing tumor growth and metastasis (Rimawi et al., 2018).

Lung cancer

EGFR-mutant NSCLC

Drugs such as erlotinib and osimertinib inhibit EGFR signaling, which is effective in EGFR-mutant tumors (Laface et al., 2023).

ALK/ROS1 rearrangements in NSCLC

Crizotinib and alectinib block ALK and ROS1 fusion proteins, preventing tumor progression (Ye et al., 2016).

Melanoma

Targeting the BRAF V600E mutation with inhibitors such as vemurafenib and dabrafenib disrupts the MAPK signaling pathway, which is critical for melanoma cell survival (Proietti et al., 2020).

Colorectal cancer

EGFR inhibitors

Drugs such as cetuximab and panitumumab block EGFR signaling in RAS wild-type tumors (Chan et al., 2017).

VEGF inhibitors

Bevacizumab reduces angiogenesis, cutting off the blood supply to tumors (Pavlidis and Pavlidis, 2013).

Chronic myeloid leukemia

Tyrosine kinase inhibitors

Drugs such as imatinib specifically inhibit the BCR–ABL fusion protein, a hallmark of CML (An et al., 2010).

Ovarian cancer

PARP inhibitors

Drugs such as olaparib and niraparib impair DNA repair in BRCA-mutated ovarian cancer, inducing cancer cell death (Kim and Nam, 2022; Ledermann, 2016).

Hepatocellular carcinoma

Multikinase inhibitors, such as sorafenib and lenvatinib, target angiogenesis and tumor proliferation pathways (Zhu et al., 2020).

Hematological malignancies

CD20-positive B-cell malignancies

Rituximab and obinutuzumab target CD20 antigens on B cells, enhancing immune-mediated cell destruction (Illidge et al., 2015; Salles et al., 2017).

Prostate cancer

Androgen receptor inhibitors

Drugs such as enzalutamide block androgen signaling, which is crucial for prostate cancer cell growth (Chen et al., 2022).

Gastrointestinal stromal tumors

Imatinib targets KIT and PDGFRA (Platelet-Derived Growth Factor Receptor Alpha) mutations, which drive gastrointestinal stromal tumor progression (Balachandran and DeMatteo, 2013).

Limitation of targeted therapy in cancer research (National Cancer Institute, n.d.)

Development of resistance

Cancer cells can develop resistance to targeted therapy. Resistance occurs when the target changes, making the therapy ineffective. Cancer cells may also find alternative growth pathways that bypass the target.

Limited effectiveness as monotherapy

Targeted therapy is often more effective when combined with other treatments, such as chemotherapy or radiation. Using multiple types of targeted therapy can also enhance effectiveness.

Challenges in drug development

Some targets are difficult to develop drugs for due to their complex structure or critical functions within the cell. These challenges limit the availability of targeted drugs for certain cancers.

Conclusion

Targeted therapy in cancer works by precisely targeting the molecular and genetic abnormalities that drive cancer growth. These therapies offer a more focused and often less toxic alternative to traditional treatments, providing hope for improved outcomes in many cancer types. The mechanisms—ranging from inhibiting growth signals, inducing apoptosis, blocking angiogenesis, delivering cytotoxic agents, to blocking hormone signaling—demonstrate the diverse and adaptable nature of targeted therapy in oncology.

References

Genomic Profiling Offers a Personalized Approach to Breast Cancer Treatment, 2024. https://www.targetedonc.com/view/genomic-profiling-offers-a-personalized-approach-to- breast-cancer-treatment.

An, X., Tiwari, A.K., Sun, Y., Ding, P.-R., Ashby, C.R., Chen, Z.-S., 2010. BCR-ABL tyrosine kinase inhibitors in the treatment of Philadelphia chromosome positive chronic myeloid leukemia: a review. Leuk. Res. 0145-2126. 34 (10), 1255–1268. doi:10.1016/j.leukres.2010.04.016.

Balachandran, V.P., DeMatteo, R.P., 2013. Targeted therapy for cancer. Surg. Oncol. Clin. N. Am. 1558-5042. 22 (4), 805–821. doi:10.1016/j.soc.2013.06.001.

Bethune, G., Bethune, D., Ridgway, N., Xu, Z., 2010. Epidermal growth factor receptor (EGFR) in lung cancer: an overview and update. J. Thorac. Dis. 2077-6624. 2 (1), 48–51. Available from: http://www.jthoracdis.com/article/download/87/pdf_15.

Bradley, R., Braybrooke, J., Gray, R., Hills, R., Liu, Z., Peto, R., Davies, L., Dodwell, D., McGale, P., Pan, H., Taylor, C., Anderson, S., Gelber, R., Gianni, L., Jacot, W., Joensuu, H., Moreno-Aspitia, A., Piccart, M., Press, M., Romond, E., Slamon, D., Suman, V., Berry, R., Boddington, C., Clarke, M., Davies, C., Duane, F., Evans, V., Gay, J., Gettins, L., Godwin, J., James, S., Liu, H., MacKinnon, E., Mannu, G., McHugh, T., Morris, P., Read, S., Straiton, E., Wang, Y., Crown, J., de Azambuja, E., Delaloge, S., Fung, H., Geyer, C., Spielmann, M., Valagussa, P., Albain, K., Anderson, S., Arriagada, R., Bartlett, J., Bergsten-Nordström, E., Bliss, J., Brain, E., Carey, L., Coleman, R., Cuzick, J., Davidson, N., Del Mastro, L., Di Leo, A., Dignam, J., Dowsett, M., Ejlertsen, B., Francis, P., Gnant, M., Goetz, M., Goodwin, P., Halpin-Murphy, P., Hayes, D., Hill, C., Jagsi, R., Janni, W., Loibl, S., Mamounas, E.P., Martín, M., Mukai, H., Nekljudova, V., Norton, L., Ohashi, Y., Pierce, L., Poortmans, P., Raina, V., Rea, D., Regan, M., Robertson, J., Rutgers, E., Spanic, T., Sparano, J., Steger, G., Tang, G., Toi, M., Tutt, A., Viale, G., Wang, X., Whelan, T., Wilcken, N., Wolmark, N., Cameron, D., Bergh, J., Pritchard, K.I., Swain, S.M., 2021. Trastuzumab for early-stage, HER2-positive breast cancer: a meta-analysis of 13 864 women in seven randomised trials. Lancet Oncol. 1470-2045. 22 (8), 1139–1150. doi:10.1016/s1470-2045(21)00288-6.

Cancer Treatment Centers of America, 2024. Common targeted therapy side effects and how to manage them. Available from: https://www.cancercenter.com/treatment-options/precision-medicine/targeted-therapy/common-side-effects.

Chan, D.L.H., Segelov, E., Wong, R.S.H., Smith, A., Herbertson, R.A., Li, B.T., Tebbutt, N., Price, T., Pavlakis, N., 2017. Epidermal growth factor receptor (EGFR) inhibitors for metastatic colorectal cancer. Cochrane Database Syst. Rev. 1469-493X. 2017 (6). Available from: http://as.wiley.com/WileyCDA/Brand/id-6.html. doi:10.1002/14651858.CD007047.pub2.

Chapman, P.B., Hauschild, A., Robert, C., Haanen, J.B., Ascierto, P., Larkin, J., Dummer, R., Garbe, C., Testori, A., Maio, M., Hogg, D., Lorigan, P., Lebbe, C., Jouary, T., Schadendorf, D., Ribas, A., O'Day, S.J., Sosman, J.A., Kirkwood, J.M., Eggermont, A.M.M., Dreno, B., Nolop, K., Li, J., Nelson, B., Hou, J., Lee, R.J., Flaherty, K.T., McArthur, G.A., 2011. Improved survival with vemurafenib in melanoma with BRAF V600E mutation. N. Engl. J. Med. 1533-4406. 364 (26), 2507–2516. Available from: http://www.nejm.org/doi/pdf/10.1056/NEJMoa1103782. doi:10.1056/NEJMoa1103782.

Chaput, G., Sumar, N., 2022. Endocrine therapies for breast and prostate cancers Essentials for primary care. Can. Fam. Physician. 0008-350X. 68 (4), 271–276. Available from: https://www.cfp.ca/content/68/4/271. doi:10.46747/cfp.6804271.

Chen, Y., Zhou, Q., Hankey, W., Fang, X., Yuan, F., 2022. Second generation androgen receptor antagonists and challenges in prostate cancer treatment. Cell Death Dis. 2041-4889. 13 (7), 632. doi:10.1038/s41419-022-05084-1.

Clinical Trials Using Gemtuzumab Ozogamicin, n.d. https://www.cancer.gov/research/participate/clinical-trials/intervention/gemtuzumab- ozogamicin.

Collins, I., Workman, P., 2006. New approaches to molecular cancer therapeutics. Nat. Chem. Biol. 1552-4450. 2 (12), 689–700. doi:10.1038/nchembio840.

Committee on Policy Issues in the Clinical Development and Use of Biomarkers for Molecularly Targeted Therapies, n.d. Biomarker Tests for Molecularly Targeted Therapies: Key to Unlocking Precision Medicine. National Academies Press.

Cuzick, J., Sestak, I., Baum, M., Buzdar, A., Howell, A., Dowsett, M., Forbes, J.F., 2010. Effect of anastrozole and tamoxifen as adjuvant treatment for early-stage breast cancer: 10 year analysis of the ATAC trial. Lancet Oncol. 1470-2045. 11 (12), 1135–1141. doi:10.1016/s1470-2045(10)70257-6.

Díaz-Rubio, E., Schmoll, H.-J., 2005. Introduction. Oncology. 0030-2414. 69 (3), 1–3. doi:10.1159/000088477.

Druker, B.J., Guilhot, F., O'Brien, S.G., Gathmann, I., Kantarjian, H., Gattermann, N., Deininger, M.W.N., Silver, R.T., Goldman, J.M., Stone, R.M., Cervantes, F., Hochhaus, A., Powell, B.L., Gabrilove, J.L., Rousselot, P., Reiffers, J., Cornelissen, J.J., Hughes, T., Agis, H., Fischer, T., Verhoef, G., Shepherd, J., Saglio, G., Gratwohl, A., Nielsen, J.L., Radich, J.P., Simonsson, B., Taylor, K., Baccarani, M., So, C., Letvak, L., Larson, R.A., 2006. Five-year follow-up of patients receiving imatinib for chronic myeloid leukemia. N. Engl. J. Med. 1533-4406. 355 (23), 2408–2417. Available from: http://content.nejm.org/cgi/reprint/355/23/2408.pdf. doi:10.1056/NEJMoa062867.

Druker, B.J., Talpaz, M., Resta, D.J., Peng, B., Buchdunger, E., Ford, J.M., Lydon, N.B., Kantarjian, H., Capdeville, R., Ohno-Jones, S., Sawyers, C.L., 2001. Efficacy and safety of a specific inhibitor of the BCR-ABL tyrosine kinase in chronic myeloid leukemia. N. Engl. J. Med. 0028-4793. 344 (14), 1031–1037. doi:10.1056/NEJM200104053441401.

Farrar, M.C., Jacobs, T.F., 2023. Tamoxifen. StatPearls Publishing, Treasure Island (FL).
Fernald, K., Kurokawa, M., 2013. Evading apoptosis in cancer. Trends Cell Biol. 0962-8924. 23 (12), 620–633. doi:10.1016/j.tcb.2013.07.006.
Ferrara, N., Hillan, K.J., Gerber, H.P., Novotny, W., 2004. Discovery and development of bevacizumab, an anti-VEGF antibody for treating cancer. Nat. Rev. Drug Discov. 1474-1776. 3 (5), 391–400. Available from: http://www.nature.com/nrd/index.html. doi:10.1038/nrd1381.
Fisher, B., Costantino, J.P., Wickerham, D.L., Cecchini, R.S., Cronin, W.M., Robidoux, A., Bevers, T.B., Kavanah, M.T., Atkins, J.N., Margolese, R.G., Runowicz, C.D., James, J.M., Ford, L.G., Wolmark, N., 2005. Tamoxifen for the prevention of breast cancer: current status of the National Surgical Adjuvant Breast and Bowel Project P-1 study. J. Natl. Cancer Inst. 1460-2105. 97 (22), 1652–1662. doi:10.1093/jnci/dji372.
Fisher, B., Costantino, J.P., Wickerham, D.L., Redmond, C.K., Kavanah, M., Cronin, W.M., Vogel, V., Robidoux, A., Dimitrov, N., Atkins, J., Daly, M., Wieand, S., Tan-Chiu, E., Ford, L., Wolmark, N., 1998. Tamoxifen for prevention of breast cancer: report of the National Surgical Adjuvant Breast and Bowel Project P-1 study. J. Natl. Cancer Inst. 0027-8874. 90 (18), 1371–1388. Available from: http://jnci.oxfordjournals.org/. doi:10.1093/jnci/90.18.1371.
Fong, P.C., Boss, D.S., Yap, T.A., Tutt, A., Wu, P., Mergui-Roelvink, M., Mortimer, P., Swaisland, H., Lau, A., O'Connor, M.J., Ashworth, A., Carmichael, J., Kaye, S.B., Schellens, J.H.M., De Bono, J.S., 2009. Inhibition of poly(ADP-ribose) polymerase in tumors from BRCA mutation carriers. N. Engl. J. Med. 1533-4406. 361 (2), 123–134. Available from: http://content.nejm.org/cgi/reprint/361/2/123.pdf. doi:10.1056/NEJMoa0900212.
Gerriets, V., Kasi, A., 2023. Bevacizumab. StatPearls Publishing, Treasure Island (FL).
Hanahan, D., Weinberg, R.A., 2011. Hallmarks of cancer: the next generation. Cell. 0092-8674. 144 (5), 646–674. doi:10.1016/j.cell.2011.02.013.
Hurwitz, H., Fehrenbacher, L., Novotny, W., Cartwright, T., Hainsworth, J., Heim, W., Berlin, J., Baron, A., Griffing, S., Holmgren, E., Ferrara, N., Fyfe, G., Rogers, B., Ross, R., Kabbinavar, F., 2004. Bevacizumab plus irinotecan, fluorouracil, and leucovorin for metastatic colorectal cancer. N. Engl. J. Med. 0028-4793. 350 (23), 2335–2342. doi:10.1056/NEJMoa032691.
Illidge, T., Klein, C., Sehn, L.H., Davies, A., Salles, G., Cartron, G., 2015. Obinutuzumab in hematologic malignancies: lessons learned to date. Cancer Treat. Rev. 0305-7372. 41 (9), 784–792. doi:10.1016/j.ctrv.2015.07.003.
Joo, W.D., Visintin, I., Mor, G., 2013. Targeted cancer therapy—are the days of systemic chemotherapy numbered? Maturitas. 1873-4111. 76 (4), 308–314. doi:10.1016/j.maturitas.2013.09.008.
Kim, D., Nam, H.J., 2022. PARP inhibitors: clinical limitations and recent attempts to overcome them. Int. J. Mol. Sci. 1422-0067. 23 (15). Available from: http://www.mdpi.com/journal/ijms. doi:10.3390/ijms23158412.
Kwok, G., Yau, T.C.C., Chiu, J.W., Tse, E., Kwong, Y.L., 2016. Pembrolizumab (Keytruda). Hum. Vaccin. Immunother. 2164-554X. 12 (11), 2777–2789. Available from: http://www.tandfonline.com/loi/khvi20. doi:10.1080/21645515.2016.1199310.
Laface, C., Maselli, F.M., Santoro, A.N., Iaia, M.L., Ambrogio, F., Laterza, M., Guarini, C., De Santis, P., Perrone, M., Fedele, P., 2023. The resistance to EGFR-TKIs in non-small cell lung cancer: from molecular mechanisms to clinical application of new therapeutic strategies. Pharmaceutics. 1999-4923. 15 (6), 1604. doi:10.3390/pharmaceutics15061604.
Ledermann, J.A., 2016. PARP inhibitors in ovarian cancer. Ann. Oncol. 0923-7534. 27, i40. doi:10.1093/annonc/mdw094.
Li, J., Wang, Q., Xia, G., Adilijiang, N., Li, Y., Hou, Z., Fan, Z., Li, J., 2023. Recent advances in targeted drug delivery strategy for enhancing oncotherapy. Pharmaceutics. 1999-4923. 15 (9), 2233. doi:10.3390/pharmaceutics15092233.
Lo Nigro, C., Macagno, M., Sangiolo, D., Bertolaccini, L., Aglietta, M., Merlano, M.C., 2019. NK-mediated antibody-dependent cell-mediated cytotoxicity in solid tumors: biological evidence and clinical perspectives. Ann. Transl. Med. 2305-5839. 7 (5), 105. doi:10.21037/atm.2019.01.42.
Malone, E.R., Oliva, M., Sabatini, P.J.B., Stockley, T.L., Siu, L.L., 2020. Molecular profiling for precision cancer therapies. Genome Med. 1756-994X. 12 (1), 8. Available from: http://www.genomemedicine.com/. doi:10.1186/s13073-019-0703-1.
Markowitz, S.D., Bertagnolli, M.M., 2009. Molecular basis of colorectal cancer. N. Engl. J. Med. 1533-4406. 361 (25), 2449. Available from: http://content.nejm.org/cgi/reprint/361/25/2449.pdf. doi:10.1056/NEJMra0804588.
Min, H.Y., Lee, H.Y., 2022. Molecular targeted therapy for anticancer treatment. Exp. Mol. Med. 2092-6413. 54 (10), 1670–1694. Available from: https://www.nature.com/emm/. doi:10.1038/s12276-022-00864-3.
National Cancer Institute, 2024. Angiogenesis inhibitors. Available from: https://www.cancer.gov/about-cancer/treatment/types/immunotherapy/angiogenesis-inhibitors-fact-sheet.

National Cancer Institute, n.d. Immune Checkpoint Inhibitors. Available from: https://www.cancer.gov/about-cancer/treatment/types/immunotherapy/checkpoint-inhibitors.

National Cancer Institute, n.d. Targeted Therapy to Treat Cancer. Available from: https://www.cancer.gov/about-cancer/treatment/types/targeted-therapies.

Nunes, A.T., Annunziata, C.M., 2017. Proteasome inhibitors: structure and function. Semin. Oncol. 1532-8708. 44 (6), 377–380. Available from: http://www.elsevier.com/inca/publications/store/6/2/3/3/7/8/index.htt. doi: 10.1053/j.seminoncol.2018.01.004.

Pavlidis, E.T., Pavlidis, T.E., 2013. Role of bevacizumab in colorectal cancer growth and its adverse effects: a review. World J. Gastroenterol. 2219-2840. 19 (31), 5051–5060. Available from: http://www.wjgnet.com/1007-9327/pdf/v19/i31/5051.pdf. doi:10.3748/wjg.v19.i31.5051.

Pfeffer, C.M., Singh, A.T.K., 2018. Apoptosis: a target for anticancer therapy. Int. J. Mol. Sci. 1422-0067. 19 (2), 448. Available from: http://www.mdpi.com/1422-0067/19/2/448/pdf. doi:10.3390/ijms19020448.

Proietti, I., Skroza, N., Michelini, S., Mambrin, A., Balduzzi, V., Bernardini, N., Marchesiello, A., Tolino, E., Volpe, S., Maddalena, P., Di Fraia, M., Mangino, G., Romeo, G., Potenza, C., 2020. BRAF inhibitors: molecular targeting and immunomodulatory actions. Cancers. 2072-6694. 12 (7), 1823. doi:10.3390/cancers12071823.

Rimawi, M., Ferrero, J.M., De La Haba-Rodriguez, J., Poole, C., De Placido, S., Osborne, C.K., Hegg, R., Easton, V., Wohlfarth, C., Arpino, G., 2018. First-line trastuzumab plus an aromatase inhibitor, with or without pertuzumab, in human epidermal growth factor receptor 2-positive and hormone receptor-positive metastatic or locally advanced breast cancer (PERTAIN): a randomized, open-label phase II trial. J. Clin. Oncol. 1527-7755. 36 (28), 2826–2835. Available from: http://ascopubs.org/doi/pdf/10.1200/JCO.2017.76.7863. doi:10.1200/JCO.2017.76.7863.

Rose, M., Burgess, J.T., O'Byrne, K., Richard, D.J., Bolderson, E., 2020. Inhibitors, clinical relevance, mechanisms of action and tumor resistance. Front. Cell Dev. Biol. 8.

Rowe, J.M., Löwenberg, B., 2013. Gemtuzumab ozogamicin in acute myeloid leukemia: a remarkable saga about an active drug. Blood. 1528-0020. 121 (24), 4838–4841. Available from: http://www.bloodjournal.org/content/bloodjournal/121/24/4838.full.pdf. doi:10.1182/blood-2013-03-490482.

Salles, G., Barrett, M., Foà, R., Maurer, J., O'Brien, S., Valente, N., Wenger, M., Maloney, D.G., 2017. Rituximab in B-cell hematologic malignancies: a review of 20 years of clinical experience. Adv. Ther. 0741-238X. 34 (10), 2232–2273. doi:10.1007/s12325-017-0612-x.

San Miguel, J.F., Schlag, R., Khuageva, N.K., Dimopoulos, M.A., Shpilberg, O., Kropff, M., Spicka, I., Petrucci, M.T., Palumbo, A., Samoilova, O.S., Dmoszynska, A., Abdulkadyrov, K.M., Schots, R., Jiang, B., Mateos, M.-V., Anderson, K.C., Esseltine, D.L., Liu, K., Cakana, A., van de Velde, H., Richardson, P.G., 2008. Bortezomib plus melphalan and prednisone for initial treatment of multiple myeloma. N. Engl. J. Med. 0028-4793. 359 (9), 906–917. doi:10.1056/nejmoa0801479.

Sever, R., Brugge, J.S., 2015. Signal transduction in cancer. Cold Spring Harb. Perspect. Med. 2157-1422. 5 (4), a006098. doi:10.1101/cshperspect.a006098.

Shiravand, Y., Khodadadi, F., Kashani, S.M.A., Hosseini-Fard, S.R., Hosseini, S., Sadeghirad, H., Ladwa, R., O'Byrne, K., Kulasinghe, A., 2022. Immune checkpoint inhibitors in cancer therapy. Curr. Oncol. 1718-7729. 29 (5), 3044–3060. doi:10.3390/curroncol29050247.

Starling, N., Neoptolemos, J., Cunningham, D., 2006. Role of erlotinib in the management of pancreatic cancer. Ther. Clin. Risk Manag. 1176-6336. 2 (4), 435–445. doi:10.2147/tcrm.2006.2.4.435.

Ulm, M., Ramesh, A.V., McNamara, K.M., Ponnusamy, S., Sasano, H., Narayanan, R., 2019. Therapeutic advances in hormone-dependent cancers: focus on prostate, breast and ovarian cancers. Endocr. Connect. 2049-3614. 8 (2), R10. doi:10.1530/ec-18-0425.

Ye, M., Zhang, X., Li, N., Zhang, Y., Jing, P., Chang, N., Wu, J., Ren, X., Zhang, J., 2016. ALK and ROS1 as targeted therapy paradigms and clinical implications to overcome crizotinib resistance. Oncotarget. 1949-2553. 7 (11), 12289–12304. doi:10.18632/oncotarget.6935.

Zhu, X.D., Tang, Z.Y., Sun, H.C., 2020. Targeting angiogenesis for liver cancer: past, present, and future. Genes Dis. 2352-3042. 7 (3), 328–335. Available from: https://www.keaipublishing.com/en/journals/genes-and-diseases/. doi:10.1016/j.gendis.2020.03.010.

CHAPTER

2

Chemotherapy agents: Current and emerging options

Harish Kumar[a], *Anusuya Bhattacharyya*[b], *Jitender Gairola*[c], *and Phulen Sarma*[d]

[a]Department of Pharmacology, AIIMS-GUWAHATI, New Delhi, India
[b]Department of Ophthalmology, AIIMS, Guwahati, India
[c]Consultant-Centre of Excellence for Use of Drones in Medicine, AIIMS, Rishikesh, India
[d]Department of Pharmacology, AIIMS, Guwahati, India

Introduction

Chemotherapy remains a cornerstone in cancer treatment, employing cytotoxic agents to target rapidly dividing cells. Current chemotherapy regimens include alkylating agents, antimetabolites, topoisomerase inhibitors, mitotic inhibitors, and targeted therapies, each with distinct mechanisms of action. The advent of personalized medicine and the integration of targeted therapies and immunotherapy have significantly improved treatment outcomes, offering better specificity and reduced systemic toxicity. Despite these advances, chemotherapy continues to pose challenges, including drug resistance, adverse effects, and quality-of-life concerns for patients.

From a social perspective, chemotherapy has a profound impact on patients, families, and healthcare systems. While it enhances survival rates and extends life expectancy, it often comes with psychosocial burdens, such as emotional distress, financial burden and reduced productivity. In addition, disparities in access to chemotherapy persist, particularly in low- and middle-income countries, where affordability and availability of novel agents remain limited. Addressing these challenges requires a multidisciplinary approach, incorporating clinical research, patient-centered care, and policy interventions to improve treatment efficacy and reduce the societal burden of cancer.

Chemotherapy is a commonly used systemic cancer treatment that involves the use of cytotoxic drugs to target and destroy rapidly dividing malignant cells.

The following are the major classes of current chemotherapy agents, their mechanisms, clinical applications, and potential side effects.

Pharmacological Approaches in Cancer Treatment
https://doi.org/10.1016/B978-0-443-38307-6.00005-3

Classification based on the mechanism of action

Based on mechanism of action

Alkylating agents

Alkylating agents exert their cytotoxic effects by covalently attaching alkyl groups to DNA, primarily at the N7 position of guanine. This modification results in DNA damage, leading to base mispairing and strand breaks. In addition, these agents form inter- and intra-strand cross-links, preventing the separation of DNA strands during replication and transcription. The accumulation of DNA damage triggers cell cycle arrest, particularly in the G1 and S phases, and activates apoptotic pathways, leading to programmed cell death. Unlike some chemotherapeutic agents, alkylating agents are effective against both dividing and nondividing cells, making them valuable in treating slow-growing tumors. Furthermore, they can generate reactive oxygen species, causing additional cellular damage. By impairing transcription and translation, alkylating agents hinder the synthesis of essential proteins, ultimately disrupting cellular functions and promoting tumor cell death (Singh et al., 2023).

Mechanism: These agents work by adding an alkyl group to the DNA, which results in DNA damage and prevents cell division. They bind to DNA, causing cross-links, preventing replication, and triggering apoptosis. They can act on both dividing and nondividing cells (National Institute of Diabetes and Digestive and Kidney Diseases, 2012).

Examples

- Cyclophosphamide
- Cisplatin
- Carboplatin
- Melphalan
- Ifosfamide

Applications: Broad use in hematologic and solid tumors (e.g., breast, lung, ovarian cancers).

Indications: Used in lymphoma, leukemia, breast cancer, ovarian cancer, and other solid tumors.

Side effects

- Myelosuppression (low blood cell counts)
- Nausea and vomiting
- Increased risk of secondary malignancies due to DNA damage
- Cyclophosphamide and its isomer, ifosfamide, and their urinary metabolites can induce bladder wall irritation and cause hemorrhagic cystitis (Haigentz, 2009)

Antimetabolites

Mechanism: Antimetabolites are cytotoxic agents. These agents mimic natural metabolites essential for DNA and RNA synthesis, thereby interfering with nucleotide synthesis and halting cell proliferation (Lansiaux, 2011).

Examples

- Methotrexate (inhibits folic acid metabolism)
- 5-Fluorouracil (5-FU) (inhibits thymidylate synthase)
- Cytarabine (for leukemia)
- Capecitabine
- Gemcitabine

Applications: Breast, colorectal, pancreatic, and hematological malignancies.

Side effects

- Myelosuppression
- Mucositis (inflammation of the digestive tract) (Sougiannis et al., 2021)
- Diarrhea and hand-foot syndrome (specific to 5-FU and capecitabine) (Milano et al., 2008)

Antimicrotubule agents (mitotic inhibitors)

Mechanism: These agents inhibit the formation or function of microtubules, which are essential for cell division. They prevent cells from completing mitosis, leading to cell death (Mukhtar et al., 2014).

Examples

- Vincristine (from the vinca alkaloids)
- Paclitaxel (from the taxanes)

Applications

- **Paclitaxel and docetaxel:** Used to treat breast cancer, ovarian cancer, and nonsmall cell lung cancer (NSCLC).
- **Vinblastine:** Lymphoma, testicular cancer.
- **Vincristine:** Leukemia, lymphoma, and neuroblastoma.

Side effects

- Peripheral neuropathy (nerve damage)
- Neutropenia (low neutrophil count)
- Hypersensitivity reactions

Topoisomerase inhibitors

Mechanism: Topoisomerases are enzymes that relieve strain in the DNA helix during replication. Inhibiting these enzymes leads to DNA damage and cell death (Yakkala et al., 2023).

Examples

- Topoisomerase I inhibitors: irinotecan and topotecan
- Topoisomerase II inhibitors: etoposide and doxorubicin
- **Indications**: Used in lung cancer, ovarian cancer, leukemia, and testicular cancer.

Side effects

- Alopecia (hair loss)
- Myelosuppression
- Cardiotoxicity (doxorubicin)

Cytotoxic antibiotics

Mechanism: These are natural or synthetic compounds that inhibit DNA and RNA synthesis. They work by intercalating into DNA strands and disrupting cell function (Reuvers et al., 2020).

Examples

- Doxorubicin (Yakkala et al., 2023)
- Bleomycin
- Daunorubicin

Indications: Used in lymphomas, leukemia, breast cancer, and sarcomas.

Side effects

- Pulmonary fibrosis (bleomycin)
- Cardiotoxicity (doxorubicin)
- Myelosuppression

Corticosteroids

- **Mechanism:** Corticosteroids have immunosuppressive effects and are directly toxic to some cancer cells. They are widely used in hematological malignancies (leukemia, lymphoma) and as supportive therapy to reduce inflammation and side effects (Faggiano et al., 2022; Inaba and Pui, 2010).

Examples and applications

- **Prednisone:** Lymphoma, leukemia.
- **Dexamethasone:** Brain tumors, multiple myeloma.

Side effects

- Hyperglycemia
- Immunosuppression
- Osteoporosis

Based on chemical structure

Alkylating agents

Structure: They contain alkyl groups, which can covalently bond to DNA.
Examples: Cyclophosphamide, ifosfamide, melphalan, busulfan

Platinum-based compounds

Structure: They contain platinum as a central atom, which forms covalent bonds with DNA (Zhang et al., 2022).

Examples

- Cisplatin
- Carboplatin
- Oxaliplatin

Indications: Primarily used in testicular cancer, ovarian cancer, lung cancer, and colorectal cancer.

Vinca alkaloids

Structure: Derived from the periwinkle plant, these agents prevent the formation of the mitotic spindle (Newton, 2006).

Examples

- Vincristine
- Vinblastine

Indications: Used in leukemia, lymphoma, and solid tumors.

Taxanes

Structure: Derived from the Pacific yew tree, these agents stabilize microtubules and prevent their disassembly (Pucci et al., 2018; Balasubramanian et al., 1998).

Examples

- Paclitaxel
- Docetaxel

Indications: Used in breast cancer, ovarian cancer, and NSCLC.

Antimetabolites

Structure: Mimic normal metabolites involved in DNA and RNA synthesis (Lansiaux, 2011; Sougiannis et al., 2021; Milano et al., 2008).

Examples

- Methotrexate
- 5-FU
- Mercaptopurine

Indications: Used in leukemias, lymphomas, and solid tumors.

Other special categories

Hormonal therapy

Mechanism: These agents block or interfere with the hormones that stimulate cancer cell growth (Bodmer and Castiglione-Gertsch, 2011).

Examples

- Tamoxifen (for estrogen receptor-positive breast cancer)
- Letrozole (for breast cancer)

Indications: Used in hormone-sensitive cancers, such as breast cancer and prostate cancer.

Immunomodulatory agents

Immunomodulatory agents (IMiDs) are drugs that enhance or suppress immune responses to target cancer cells. They act by:

Mechanism: Immunomodulator **agents** (IMiDs) enhance the body's immune response, aiding in the recognition and elimination of cancer cells. These agents have become a key component in the treatment of multiple myeloma, significantly contributing to its therapeutic advancement. Pomalidomide, the most recent IMiD introduced into clinical practice, is widely used for its efficacy in managing the disease (Minařík and Ševčíková, 2022).

Modulating cytokine production (e.g., reducing proinflammatory cytokines).
Enhancing T-cell and NK-cell activity, boosting antitumor immunity.
Inhibiting angiogenesis, restricting tumor growth.

Examples

- **Thalidomide**: multiple myeloma, reduces TNF-α and angiogenesis
- **Lenalidomide**: multiple myeloma, boosts T- and NK-cell function
- **Pomalidomide**: multiple myeloma, more potent than lenalidomide
- **Interleukins (IL-2, IL-12)**: stimulate immune cell proliferation
- **Interferons (IFN-α, IFN-β)**: enhance immune surveillance

Indications: Multiple myeloma and other hematologic cancers (Cruz, 2016).
Side effects (Cruz, 2016)
Immunomodulatory agents in cancer treatment
Hematological: anemia, neutropenia, thrombocytopenia
Gastrointestinal: diarrhea, nausea, constipation
Systemic: fatigue, fever, peripheral edema
Neurological: tremor, headache, muscle cramps
Cardiovascular: hypertension, thromboembolism
Respiratory: dyspnea, cough, infections
Dermatological: rash, pruritus
Psychiatric: insomnia, anxiety
Immune-related: autoimmune reactions, increased infection risk

Other newer or investigational categories

Targeted therapy (small-molecule inhibitors) (Liu et al., 2022)

Targeted therapy offers better selectivity, efficacy, and tolerability than traditional treatments. Small-molecule inhibitors, known for their broad targets, oral convenience, and central Nervous System (CNS) penetration, are widely used. The side effects caused by nonspecific targets are significantly minimized (Liu et al., 2022).

Mechanism: Target specific molecular targets, such as proteins or enzymes involved in cancer cell growth and survival.

Examples

- Imatinib (for Chronic Myeloid Leukemia (CML))
- Gefitinib (for epidermal growth factor receptor-mutant lung cancer)

Monoclonal antibodies

Monoclonal antibodies (mAbs) have become key therapeutic agents for oncology, offering superior selectivity, specificity, and immune activation compared to traditional anticancer drugs. Advances in antibody engineering are driving the development of next-generation formats with enhanced pharmacokinetics, selectivity, and efficacy, showing promising clinical outcomes (Zhu and Yan, 2011). mAb-based immunotherapy has become a fundamental component of cancer treatment, complementing surgery, radiation, and chemotherapy. These antibodies employ diverse mechanisms of action with significant clinical relevance (Zahavi and Weiner, 2020).

Mechanism: These are antibodies that bind to specific targets on cancer cells and block their growth or mark them for destruction (Zahavi and Weiner, 2020).

Examples

- Rituximab (for lymphoma)
- Trastuzumab (for human epidermal growth factor receptor 2-positive breast cancer)

Emerging chemotherapy options

Advancements in oncology research have introduced new chemotherapy agents designed to enhance efficacy and minimize toxicity. Immunotherapies, such as checkpoint inhibitors and adoptive cell therapies, have become effective approaches for combating drug resistance in cancer. By leveraging the body's immune system to identify and destroy cancer cells, these therapies introduce innovative mechanisms capable of bypassing conventional drug resistance (Garg et al., 2024).

Emerging chemotherapy options are given in the following.

Antibody–drug conjugates

Antibody–drug conjugates (ADCs) are mAbs attached to cytotoxic drugs, enabling precise targeting and effective cancer cell destruction. Since the FDA approved the first ADC, Mylotarg (gemtuzumab ozogamicin), in 2000, 14 ADCs have gained global market approval, with over 100 candidates in clinical trials. Known as "biological missiles," ADCs are revolutionizing targeted cancer therapy (Fu et al., 2022; Riccardi et al., 2023).

- **Mechanism:** mAbs linked to cytotoxic agents, selectively targeting tumor cells.

Examples

- **Trastuzumab deruxtecan (Enhertu)**: HER2+ breast and gastric cancer
- **Sacituzumab govitecan (Trodelvy)**: Triple-negative breast cancer

- **Polatuzumab vedotin**: Diffuse large B-cell lymphoma (Fu et al., 2022)
- **Advantages:** Increased selectivity, reduced off-target effects

Poly (ADP-ribose) polymerase inhibitors

Poly (ADP-ribose) polymerase (PARP) inhibitors treat breast and ovarian cancers by disrupting DNA repair mechanisms, making them a promising approach in cancer therapy (Dilmac and Ozpolat, 2023).

Mechanism: PARP inhibitors hinder the activity of the PARP enzyme, which plays a critical role in DNA repair by detecting and repairing single-strand DNA breaks. In cancer cells with defective homologous recombination repair (e.g., BRCA (BReast CAncer gene)1/2 mutations), this leads to the accumulation of DNA damage and cell death (synthetic lethality). They block DNA repair, leading to apoptosis in tumor cells with BRCA mutations (Dilmac and Ozpolat, 2023).

Examples

- Olaparib
- Rucaparib
- Niraparib

Applications: Ovarian, breast, and prostate cancers.

DNA damage response inhibitors

DNA damage response inhibitors (DDRi) are a class of targeted therapies that interfere with the cellular machinery responsible for detecting and repairing DNA damage. By impairing DNA repair pathways, DDRi increase genomic instability in cancer cells, leading to cell cycle arrest, apoptosis, and enhanced sensitivity to chemotherapy or radiotherapy (Jurkovicova et al., 2022).

- **Mechanism:** Inhibit proteins involved in DNA repair pathways, increasing tumor susceptibility to DNA damage (Jurkovicova et al., 2022).

Ataxia telangiectasia and Rad3-related inhibitors

Ataxia telangiectasia and Rad3-related (ATR) is a serine/threonine kinase activated in response to replication stress and Single-strand breaks (SSBs). It stabilizes stalled replication forks and halts the cell cycle to allow DNA repair. ATR inhibitors block this protective mechanism, preventing cell cycle arrest and repair, thereby driving replication stress to lethal levels. This is particularly effective in tumors with preexisting DNA repair defects.

Example: Ceralasertib (Saldivar et al., 2017).

CHK1 (checkpoint kinases) inhibitors

CHK1 and CHK2 are checkpoint kinases that regulate cell cycle arrest in response to DNA damage. CHK1 prevents cells from entering mitosis with damaged DNA by arresting them at the G2/M checkpoint. CHK2 primarily regulates the G1/S checkpoint. DDRi targeting CHK1/2 disrupt these checkpoints, forcing cells with damaged DNA to enter mitosis prematurely, resulting in mitotic catastrophe and apoptosis.

Example: Prexasertib (Patil et al., 2013; Lowery et al., 2017).

Wee1 inhibitors

Wee1 is a kinase that inhibits CDK1 and regulates the G2/M checkpoint, preventing cells with damaged DNA from entering mitosis. Inhibition of Wee1 overrides this checkpoint, driving the cell into mitosis with unrepaired DNA. This causes catastrophic genomic instability and cell death, particularly in p53-deficient cancer cells, which lack proper DNA damage surveillance mechanisms (Vriend et al., 2013).

Example: Adavosertib (Falchook et al., 2023).

Applications: Ovarian, lung, and colorectal cancers.

Targeted chemotherapy nanoparticles

Nanomedicine has emerged as a groundbreaking approach in cancer therapy, overcoming the limitations of traditional chemotherapy. By utilizing the distinctive properties of nanomaterials, this strategy enhances drug selectivity, efficacy, and safety. Nanocarriers are designed to deliver therapeutic agents directly to tumor sites through either passive targeting (via the enhanced permeability and retention effect) or active targeting (through ligand–receptor interactions), thereby minimizing off-target toxicity. Moreover, these carriers can be tailored to respond to internal stimuli (e.g., pH, enzymes) or external triggers (e.g., light, temperature), enabling controlled and sustained drug release. This targeted delivery system boosts therapeutic effectiveness while reducing adverse effects. Recent progress in nanomedicine includes the development of advanced drug delivery platforms such as liposomes, dendrimers, and polymeric nanoparticles, which enhance pharmacokinetics and bioavailability. However, challenges such as large-scale production, stability, and potential toxicity of nanomedications continue to hinder their widespread clinical application. Overcoming these hurdles is essential for realizing the full potential of nanomedicine in cancer therapy (Arafat et al., 2024).

Mechanism: Use of nanoparticles to deliver chemotherapeutic drugs specifically to tumor tissues, reducing systemic toxicity.

Examples

- **Doxil** (liposomal doxorubicin)
- **Abraxane (albumin-bound paclitaxel) applications:** Breast, ovarian, and pancreatic cancers

Tumor-activated prodrugs

Tumor-activated prodrugs, like evofosfamide (TH-302), are inactive until metabolized by tumor-specific enzymes, minimizing systemic toxicity and selectively targeting hypoxic tumor cells, showing promise in pancreatic and soft tissue cancers (Grande et al., 2021).

Mechanism: Inactive until metabolized by tumor-specific enzymes, minimizing systemic toxicity.

Examples

Evofosfamide (TH-302): activated under hypoxic tumor conditions.

- **Applications:** Pancreatic and soft tissue cancers (Grande et al., 2021).

Chemotherapy resistance and overcoming it

Emerging therapies are tackling the challenge of chemotherapy resistance, which continues to be a major hurdle in cancer treatment. Cancer cells acquire resistance through multiple mechanisms, such as genetic mutations, increased expression of efflux pumps, alterations in signaling pathways, and changes in the tumor microenvironment. To overcome this, current strategies emphasize the use of targeted therapies, combination regimens, immunotherapies, and precision medicine. Combining traditional cytotoxic agents with molecularly targeted inhibitors has demonstrated synergistic effects, helping to bypass resistance pathways. Immunotherapeutic approaches, such as immune checkpoint inhibitors and adoptive cell therapies, strengthen the antitumor immune response, and reduce immune evasion. Furthermore, biomarker-driven strategies are increasingly being employed to predict and monitor treatment efficacy, enabling personalized and optimized therapies. These innovative approaches show promise in enhancing patient outcomes by overcoming drug resistance and shaping the future of cancer management.

- **Multidrug resistance modulators:** Agents like tariquidar block efflux pumps (P–gp) to retain chemotherapy inside tumor cells (Waghray and Zhang, 2018).
- **Epigenetic modifiers:** DNA methyltransferase inhibitors and histone deacetylase inhibitors are being studied to overcome resistance (Yu et al., 2024).
- **Combination therapies:** Combining chemotherapy with immune checkpoint inhibitors (e.g., pembrolizumab) or targeted therapy agents (Li et al., 2023).

Future directions

- **Personalized chemotherapy:** Genomic profiling to customize drug regimens based on individual tumor biology (Tsimberidou et al., 2020).
- **Next-Gen ADCs:** With enhanced stability and targeting capacity (Fu et al., 2022).
- **Tumor microenvironment-targeted agents:** Drugs altering the tumor microenvironment to enhance chemotherapy efficacy (Xiao and Yu, 2021).

Summary

Chemotherapy remains vital in cancer treatment, using cytotoxic agents such as alkylating agents, antimetabolites, topoisomerase inhibitors, and microtubule inhibitors. However, their nonspecific toxicity limits effectiveness. Emerging options, such as ADCs, DDRi, and nanoparticle-based delivery systems, offer enhanced specificity and reduced side effects. Immunochemotherapy combines chemotherapy with immune checkpoint inhibitors, boosting antitumor immunity.

To optimize chemotherapy outcomes, a personalized approach based on tumor genetics and molecular profiling is essential. This can optimize outcomes and overcome resistance. Combining traditional cytotoxic agents with novel targeted therapies and immunomodulators enhances efficacy while minimizing toxicity. Further research into biomarker-driven selection of chemotherapy agents and resistance mechanisms will refine therapeutic strategies and improve patient survival rates.

References

Arafat, M., Sakkal, M., Beiram, R., AbuRuz, S., 2024. Nanomedicines: emerging platforms in smart chemotherapy treatment—a recent review. Pharmaceuticals. 1424-8247. 17 (3), 315. doi:10.3390/ph17030315.

Balasubramanian, B.N., Kadow, J.F., Kramer, R.A., Vyas, D.M., 1998. Recent developments in cancer cytotoxics. In: Annual Reports in Medicinal Chemistry, vol. 33. Academic Press Inc., United States, pp. 151–162. (Chapter 15). Available from: http://www.elsevier.com/wps/find/bookdescription.cws_home/BS_ARMC/description#description. doi:10.1016/S0065-7743(08)61080-5.

Bodmer, A., Castiglione-Gertsch, M., 2011. Role of hormonal manipulations in patients with hormone-sensitive metastatic breast cancer. Eur. J. Cancer 47 (Suppl 3), S28–S37. doi:10.1016/S0959-8049(11)70144-7.

Cruz, M.P., 2016. Lenalidomide (revlimid): a thalidomide analogue in combination with dexamethasone for the treatment of all patients with multiple myeloma. P T. 41, 308–313.

Dilmac, S., Ozpolat, B., 2023. Mechanisms of PARP-inhibitor-resistance in BRCA-mutated breast cancer and new therapeutic approaches. Cancers. 2072-6694. 15 (14), 3642. doi:10.3390/cancers15143642.

Faggiano, A., Mazzilli, R., Natalicchio, A., Adinolfi, V., Argentiero, A., Danesi, R., D'Oronzo, S., Fogli, S., Gallo, M., Giuffrida, D., Gori, S., Montagnani, M., Ragni, A., Renzelli, V., Russo, A., Silvestris, N., Franchina, T., Tuveri, E., Cinieri, S., Colao, A., Giorgino, F., Zatelli, M.C., 2022. Corticosteroids in oncology: use, overuse, indications, contraindications. An Italian Association of Medical Oncology (AIOM)/ Italian Association of Medical Diabetologists (AMD)/ Italian Society of Endocrinology (SIE)/ Italian Society of Pharmacology (SIF) multidisciplinary consensus position paper. Crit. Rev. Oncol. Hematol. 1040-8428. 180, 103826. doi:10.1016/j.critrevonc.2022.103826.

Falchook, G.S., Sachdev, J., Imedio, E.R., Kumar, S., Mugundu, G.M., Jenkins, S., Chmielecki, J., Jones, S., Spigel, D.R., Johnson, M., 2023. A phase Ib study of adavosertib, a selective Wee1 inhibitor, in patients with locally advanced or metastatic solid tumors. Invest. New Drugs. 1573-0646. 41 (3), 493–502. Available from: https://www.springer.com/journal/10637. doi:10.1007/s10637-023-01371-6.

Fu, Z., Li, S., Han, S., Shi, C., Zhang, Y., 2022. Antibody drug conjugate: the "biological missile" for targeted cancer therapy. Signal Transduct. Target. Ther. 2059-3635. 7 (1), 93. doi:10.1038/s41392-022-00947-7.

Garg, P., Malhotra, J., Kulkarni, P., Horne, D., Salgia, R., Singhal, S.S., 2024. Emerging therapeutic strategies to overcome drug resistance in cancer cells. Cancers. 2072-6694. 16 (13), 2478. doi:10.3390/cancers16132478.

Grande, E., Rodriguez-Antona, C., López, C., Alonso-Gordoa, T., Benavent, M., Capdevila, J., Teulé, A., Custodio, A., Sevilla, I., Hernando, J., Gajate, P., Molina-Cerrillo, J., José Díez, J., Santos, M., Lanillos, J., García-Carbonero, R., 2021. Sunitinib and evofosfamide (TH −302) in systemic treatment-naïve patients with grade 1/2 metastatic pancreatic neuroendocrine tumors: the GETNE −1408 trial. Oncologist. 1083-7159. 26 (11), 941–949. doi:10.1002/onco.13885.

Haigentz, M., 2009. In: Eisele, D.W., Smith, R.V. (Eds.), Complications in Head and Neck Surgery. 2nd ed. Philadelphia, PA: Mosby, pp. 181–188.

Inaba, H., Pui, C.H., 2010. Glucocorticoid use in acute lymphoblastic leukaemia. Lancet Oncol. 1470-2045. 11 (11), 1096–1106. doi:10.1016/S1470-2045(10)70114-5.

Jurkovicova, D., Neophytou, C.M., Gašparović, A.Č., Gonçalves, A.C., 2022. DNA damage response in cancer therapy and resistance: challenges and opportunities. Int. J. Mol. Sci. 1422-0067. 23 (23), 14672. Available from: http://www.mdpi.com/journal/ijms. doi:10.3390/ijms232314672.

Lansiaux, A., 2011. Les antimétabolites. Bull. Cancer. 1769-6917. 98 (11), 1263–1274. Available from: http://www.jle.com/e-docs/00/04/6E/83/vers_alt/VersionPDF.pdf. doi:10.1684/bdc.2011.1476.

Li, B., Jin, J., Guo, D., Tao, Z., Hu, X., 2023. Immune checkpoint inhibitors combined with targeted therapy: the recent advances and future potentials. Cancers. 2072-6694. 15 (10), 2858. doi:10.3390/cancers15102858.

Liu, G.H., Chen, T., Zhang, X., Ma, X.L., Shi, H.S., 2022. Small molecule inhibitors targeting the cancers. MedComm. 2688-2663. 3 (4), e181. Available from: http://onlinelibrary.wiley.com/journal/26882663. doi:10.1002/mco2.181.

Lowery, C.D., VanWye, A.B., Dowless, M., Blosser, W., Falcon, B.L., Stewart, J., Stephens, J., Beckmann, R.P., Lin, A.B., Stancato, L.F., 2017. The checkpoint kinase 1 inhibitor prexasertib induces regression of preclinical models of human neuroblastoma. Clin. Cancer Res. 1557-3265. 23 (15), 4354–4363. Available from: http://clincancerres.aacrjournals.org/content/23/15/4354.full-text.pdf. doi:10.1158/1078-0432.CCR-16-2876.

Milano, G., Etienne-Grimaldi, M.-C., Mari, M., Lassalle, S., Formento, J.-L., Francoual, M., Lacour, J.-P., Hofman, P., 2008. Candidate mechanisms for capecitabine-related hand–foot syndrome. Br. J. Clin. Pharmacol. 0306-5251. 66 (1), 88–95. doi:10.1111/j.1365-2125.2008.03159.x.

Minařík, J., Ševčíková, S., 2022. Immunomodulatory agents for multiple myeloma. Cancers. 2072-6694. 14 (23), 5759. doi:10.3390/cancers14235759.

Mukhtar, E., Adhami, V.M., Mukhtar, H., 2014. Targeting microtubules by natural agents for cancer therapy. Mol. Cancer Ther. 1538-8514. 13 (2), 275–284. Available from: http://mct.aacrjournals.org/content/13/2/275.full.pdf+html. doi:10.1158/1535-7163.MCT-13-0791.

National Institute of Diabetes and Digestive and Kidney Diseases, 2012. LiverTox®: Clinical and Research Information on Drug-Induced Liver Injury [Internet]. Bethesda (MD): National Institute of Diabetes and Digestive and Kidney Diseases. PMID: 31643176.

Newton, H.B., 2006. Clinical pharmacology of brain tumor chemotherapy. In: Handbook of Brain Tumor Chemotherapy. Elsevier Inc, United States, pp. 21–43. Available from: http://www.sciencedirect.com/science/book/9780120884100. doi:10.1016/B978-012088410-0/50040-8.

Patil, M., Pabla, N., Dong, Z., 2013. Checkpoint kinase 1 in DNA damage response and cell cycle regulation. Cell. Mol. Life Sci. 1420-682X. 70 (21), 4009–4021. doi:10.1007/s00018-013-1307-3.

Pucci, P., Rescigno, P., Sumanasuriya, S., de Bono, J., Crea, F., 2018. Hypoxia and noncoding RNAs in taxane resistance. Trends Pharmacol. Sci. 0165-6147. 39 (8), 695–709. doi:10.1016/j.tips.2018.05.002.

Reuvers, T.G.A., Kanaar, R., Nonnekens, J., 2020. DNA damage-inducing anticancer therapies: from global to precision damage. Cancers. 2072-6694. 12 (8), 1–21. Available from: https://www.mdpi.com/2072-6694/12/8/2098/pdf. doi:10.3390/cancers12082098.

Riccardi, F., Dal Bo, M., Macor, P., Toffoli, G., 2023. A comprehensive overview on antibody-drug conjugates: from the conceptualization to cancer therapy. Front. Pharmacol. 1663-9812. 14, 1274088. doi:10.3389/fphar.2023.1274088.

Saldivar, J.C., Cortez, D., Cimprich, K.A., 2017. The essential kinase ATR: ensuring faithful duplication of a challenging genome. Nat. Rev. Mol. Cell Biol. 1471-0080. 18 (10), 622–636. Available from: http://www.nature.com/molcellbio. doi:10.1038/nrm.2017.67.

Singh, A., Singh, B., Navneet, 2023. Bioactive compounds in cancer care and prevention. In: Role of Nutrigenomics in Modern-day Healthcare and Drug Discovery. Elsevier BV, pp. 439–468. doi:10.1016/b978-0-12-824412-8.00007-2.

Sougiannis, A.T., VanderVeen, B.N., Davis, J.M., Fan, D., Murphy, E.A., 2021. Understanding chemotherapy-induced intestinal mucositis and strategies to improve gut resilience. Am. J. Physiol. Gastrointest. Liver Physiol. 1522-1547. 320 (5), G712. Available from: https://journals.physiology.org/doi/10.1152/ajpgi.00380.2020. doi:10.1152/AJPGI.00380.2020.

Tsimberidou, A.M., Fountzilas, E., Nikanjam, M., Kurzrock, R., 2020. Review of precision cancer medicine: evolution of the treatment paradigm. Cancer Treat. Rev. 0305-7372. 86, 102019. doi:10.1016/j.ctrv.2020.102019.

Vriend, L.E.M., De Witt Hamer, P.C., Van Noorden, C.J.F., Würdinger, T., 2013. WEE1 inhibition and genomic instability in cancer. Biochim. Biophys. Acta Rev. Cancer. 1879-2561. 1836 (2), 227–235. doi:10.1016/j.bbcan.2013.05.002.

Waghray, D., Zhang, Q., 2018. Inhibit or evade multidrug resistance P-glycoprotein in cancer treatment. J. Med. Chem. 0022-2623. 61 (12), 5108–5121. doi:10.1021/acs.jmedchem.7b01457.

Xiao, Y., Yu, D., 2021. Tumor microenvironment as a therapeutic target in cancer. Pharmacol. Ther. 0163-7258. 221, 107753. doi:10.1016/j.pharmthera.2020.107753.

Yakkala, P.A., Penumallu, N.R., Shafi, S., Kamal, A., 2023. Prospects of topoisomerase inhibitors as promising anti-cancer agents. Pharmaceuticals. 1424-8247. 16 (10), 1456. Available from: https://www.mdpi.com/journal/pharmaceuticals. doi:10.3390/ph16101456.

Yu, X., Zhao, H., Wang, R., Chen, Y., Ouyang, X., Li, W., Sun, Y., Peng, A., 2024. Cancer epigenetics: from laboratory studies and clinical trials to precision medicine. Cell Death Discov. 2058-7716. 10 (1), 28. doi:10.1038/s41420-024-01803-z.

Zahavi, D., Weiner, L., 2020. Monoclonal antibodies in cancer therapy. Antibodies. 2073-4468. 9 (3), 34. doi:10.3390/antib9030034.

Zhang, C., Xu, C., Gao, X., Yao, Q., 2022. Platinum-based drugs for cancer therapy and anti-tumor strategies. Theranostics. 1838-7640. 12 (5), 2115–2132. doi:10.7150/thno.69424.

Zhu, Z., Yan, L., 2011. Next generation of antibody therapy for cancer. Chin. J. Cancer. 1000-467X. 30 (5), 293–302. doi:10.5732/cjc.011.10123.

CHAPTER

3

Immunotherapy: Harnessing the immune system in cancer treatment

Seema Bansal[a,b], *Rishabh Chaudhary*[a], *Sowvik Bag*[a], *Phulen Sarma*[b], and *Nitin Bansal*[c]

[a]M.M. College of Pharmacy, Maharishi Markandeshwar (Deemed to be University), Ambala, Haryana, India

[b]Department of Pharmacology, Post Graduate Institute of Medical Education and Research, Chandigarh, India

[c]Department of Pharmaceutical Sciences, Chaudhary Bansi Lal University, Bhiwani, India

Abbreviations

ALL	acute lymphoblastic leukemia
APCs	antigen-presenting cells
CAR	chimeric antigen receptors
CAR-T	chimeric antigen receptor T cell
CD80/CD86	B7 family members (costimulatory molecules on antigen-presenting cells)
CRES	CAR-T cell-related encephalopathy syndrome
CRS	cytokine release syndrome
CTL	cytotoxic T lymphocyte
CTLA-4	cytotoxic T-lymphocyte-associated protein 4
DNA	deoxyribonucleic acid
GM-CSF	granulocyte–macrophage colony-stimulating factor
GVAX	granulocyte–macrophage colony-stimulating factor-expressing vaccine
HBV	hepatitis B virus
HPV	human papillomavirus
IL-2	interleukin-2
irAEs	immune-related adverse events
MDSCs	myeloid-derived suppressor cells
MHC	major histocompatibility complex
NK	natural killer (cells)
NSCLC	nonsmall cell lung cancer

Pharmacological Approaches in Cancer Treatment
https://doi.org/10.1016/B978-0-443-38307-6.00012-0

PD-1	programmed death-1
PD-L1	programmed death-ligand 1
PP2A	protein phosphatase 2A
PSA	prostate-specific antigen
ROS	reactive oxygen species
SHP-2	Src homology region 2-containing phosphatase-2
T Cells	T lymphocytes
TCR	T-cell receptor
TILs	tumor-infiltrating lymphocytes
Tregs	regulatory T cells
T-VEC	talimogene laherparepvec

Introduction

Cancer is one of the world's major causes of illness and mortality, with millions of new cases diagnosed each year. Traditional cancer therapies, such as surgery, chemotherapy, and radiation therapy, sometimes have poor success rates and can produce serious adverse effects because of their nonspecificity (Huang et al., 2012). These therapies not only target cancer cells but also damage healthy tissues, resulting in a variety of side effects and lowering the patient's quality of life. In recent years, immunotherapy has received much attention as a new and promising method for cancer treatment. Immunotherapy uses the immune system to locate and remove cancer cells, providing a focused and potentially more effective alternative to traditional medicines. The immune system can discern between normal and aberrant cells, such as cancer cells, using a complex network of cellular and molecular connections. Immunotherapy seeks to produce long-term antitumor effects by augmenting and directing natural immune responses. The idea of harnessing the immune system to combat cancer dates back to the early 20th century, but substantial advances have only been realized in the last few decades. Advances in molecular biology, genetics, and immunology have provided a better understanding of how the immune system may be regulated to treat cancer. This has led to the development of various immunotherapeutic methods, including checkpoint inhibitors, chimeric antigen receptor T-cell (CAR-T) treatment, cancer vaccinations, and oncolytic viruses (Li et al., 1999).

Checkpoint inhibitors are one of the most popular and commonly utilized immunotherapies. These medicines disrupt inhibitory pathways in T lymphocytes, allowing them to assault cancer cells more effectively. CAR-T cell treatment involves genetically modifying a patient's T cells to express receptors that selectively target cancer antigens. Cancer vaccines aim to elicit an immunological response against tumor-specific antigens (TSAs) or tumor-associated antigens (TAAs), whereas oncolytic viruses selectively infect and lyse cancer cells, resulting in immune activation. Despite immunotherapy's extraordinary success in specific cancers, several hurdles remain. These include immune-related adverse events (irAEs), tumor heterogeneity, and an immunosuppressive tumor microenvironment. Furthermore, not all patients respond to immunotherapy, highlighting the need for predictive biomarkers to identify individuals most likely to benefit from these therapies (Tay et al., 2023).

This chapter covers immunopathophysiology, molecular mechanisms, and vaccines targeting the immune system to prevent cancer.

Immunopathophysiology of cancer

Cancer pathophysiology is strongly linked to the immune system, which plays a dual function in inhibiting and supporting tumor growth. The immune system detects and eliminates abnormal cells via immune surveillance, predominantly mediated by cytotoxic T lymphocytes (CTLs), natural killer (NK) cells, and macrophages. These immune cells recognize and kill cells that contain aberrant or foreign antigens found on major histocompatibility complex (MHC) molecules. Cancer cells, on the contrary, can elude immune recognition and death via a variety of methods, which contributes to tumor growth. One such method is downregulating or deleting MHC class I molecules on tumor cell surfaces, reducing CTL recognition. Furthermore, cancer cells can generate immune checkpoint proteins like PD-L1, which bind to PD-1 receptors on T cells, suppressing T-cell function and allowing the tumor to avoid immune-mediated death. Tumors can secrete immunosuppressive cytokines such as TGF-β and IL-10, which alter the tumor microenvironment and promote the development of regulatory T cells (Tregs) and myeloid-derived suppressor cells (MDSCs), inhibiting effective antitumor immune responses (Iglesias-Escudero et al., 2023).

Cancer immunoediting explains the dynamic link between the immune system and cancer. This process comprises three stages: elimination, equilibrium, and escape. During the elimination phase, the immune system effectively recognizes and eliminates emerging tumor cells. During the equilibrium phase, the immune system inhibits tumor development but does not eliminate it, resulting in dormancy, during which tumor cells can collect new mutations. Finally, during the escape phase, tumor cells undergo genetic and epigenetic modifications that prevent immune identification and killing, resulting in uncontrolled tumor growth and metastasis. This escape is aided by the previously outlined immune evasion mechanisms, the recruitment of immunosuppressive cells, and the formation of a protumorigenic milieu.

Furthermore, persistent inflammation, commonly caused by infections, autoimmune illnesses, or environmental factors, might increase tumor formation and progression. Inflammatory cells in the tumor microenvironment, such as macrophages and neutrophils, can generate reactive oxygen species and proinflammatory cytokines, causing DNA damage and promoting tumor cell proliferation and survival. These cells can also stimulate angiogenesis, tissue remodeling, and metastasis, all of which contribute to cancer progression (Emens and Loi, 2023) (Fig. 3.1).

Molecular mechanisms of immunotherapy

Immunotherapy's success is determined by its capacity to modify the immune system to target and eradicate cancer cells. Several critical molecular pathways underlying the various forms of immunotherapies are now being utilized. These include checkpoint inhibitors, CAR-T cell therapy, cancer vaccinations, and oncolytic viruses.

Checkpoint inhibitors

Checkpoint inhibitors are monoclonal antibodies that target inhibitory mechanisms restricting T-cell activation. These pathways are critical for sustaining self-tolerance and

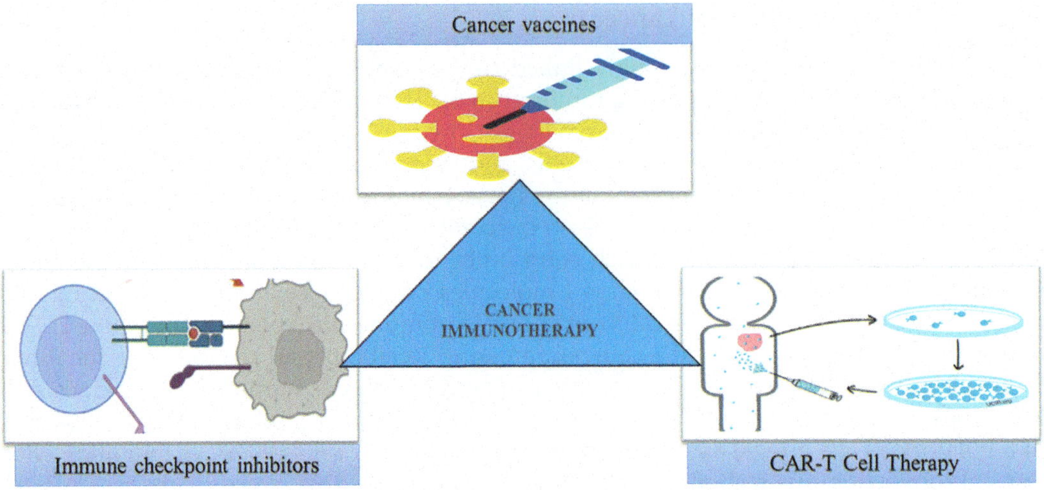

FIG. 3.1 Types of cancer immunotherapy.

controlling the length and intensity of physiological immune responses. However, tumors frequently use these mechanisms to elude immune monitoring (Li and Guo, 2023).

Cytotoxic T-lymphocyte-associated protein 4

Cytotoxic T-lymphocyte-associated protein 4 (CTLA-4) is a crucial immunological checkpoint receptor that regulates the immune system by inhibiting T-cell activation. This receptor is a member of the immunoglobulin superfamily, and it is mainly found on activated T cells and Tregs. The molecular structure of CTLA-4 is very similar to that of CD28, another T-cell receptor (TCR), with both receptors attaching to the B7 family members CD80 (B7-1) and CD86 (B7-2), which are found on antigen-presenting cells (APCs). Despite their structural similarities, CTLA-4 and CD28 have opposite activities. CD28 provides a costimulatory signal required for T-cell activation and survival, whereas CTLA-4 gives an inhibitory signal, inhibiting immunological activation, preserving immune homeostasis, and avoiding autoimmunity. CTLA-4's increased binding affinity for CD80/CD86 enables it to outcompete CD28, suppressing T-cell responses when needed. A variety of mechanisms mediate CTLA-4's inhibitory action. CTLA-4 mainly competes with CD28 to bind to CD80/CD86. Given its increased affinity, CTLA-4 effectively sequesters these ligands, preventing CD28 from providing the requisite costimulatory signals for T-cell activation. Another process is trans-endocytosis, in which CTLA-4 may remove CD80/CD86 from the surface of APCs, causing internalization and degradation, further lowering the availability of these costimulatory molecules. Furthermore, when CTLA-4 interacts with its ligands, it can attract phosphatases like SHP-2 and PP2A to the immunological synapse, dephosphorylating critical signaling molecules and blocking downstream signaling pathways (Azimnasab-sorkhabi et al., 2023).

Furthermore, CTLA-4 is constitutively expressed at high levels on Tregs, a type of T cell that helps maintain immunological tolerance. Tregs use CTLA-4 to prevent the activation of conventional T cells and promote immunological suppression, which contributes to the overall inhibitory environment.

The goal of therapeutically targeting CTLA-4 is to boost antitumor immunity by inhibiting its inhibitory role, which promotes T-cell activation and proliferation. The first immune checkpoint inhibitor licensed for cancer therapy was ipilimumab, a monoclonal antibody that inhibits CTLA-4. Ipilimumab binds to CTLA-4 and inhibits its association with CD80/CD86, removing the inhibitory signal from T cells and enabling CD28-mediated costimulatory signaling to occur unhindered. This inhibition promotes T-cell activation and a more robust antitumor immune response. Clinical studies have shown that ipilimumab has considerable advantages, notably in patients with advanced melanoma, where it has increased overall survival and resulted in persistent responses in a subgroup of patients (Obeagu and Obeagu, 2024).

However, inhibiting CTLA-4 increases immunological activation, which has certain downsides. Ipilimumab is associated with irAEs, which are side effects caused by an overactive immune system. These irAEs can vary from minor symptoms such as rash and diarrhea to severe illnesses, including autoimmune hepatitis, colitis, and endocrinopathies. Managing these irAEs necessitates close monitoring and, in some instances, the use of immunosuppressive medications such as corticosteroids.

Combination therapy is being investigated to improve the efficacy of CTLA-4 blocking while mitigating its limitations. Combining CTLA-4 inhibitors with other immunotherapeutic drugs, such as PD-1/PD-L1 inhibitors, can result in synergistic benefits since they target separate phases of the immune response. This combination has enhanced effectiveness in various malignancies, including melanoma and nonsmall cell lung cancer (NSCLC). Furthermore, combining CTLA-4 inhibitors with conventional therapies, such as chemotherapy or radiotherapy, can increase tumor antigen release and promote immune activation, converting "cold" tumors, which are poorly immunogenic, into "hot" tumors, which are more vulnerable to immune attack. In addition, combining CTLA-4 inhibition with other immunotherapies, such as cancer vaccines or oncolytic viruses, can boost antitumor immunity (Di Giacomo et al., 2024) (Fig. 3.2).

Programmed death-1/programmed death-ligand 1

Another crucial immune checkpoint is the programmed death-1 (PD-1)/programmed death-ligand 1 (PD-L1) pathway, which helps to maintain immunological homeostasis and avoid autoimmunity. PD-1 is a receptor on T-, B-, and NK-cell surfaces. PD-L1 and PD-L2 are present in various cells, including cancer cells and tumor-infiltrating immune cells. The interaction of PD-1 and its ligands generates an inhibitory signal that lowers T-cell activity, resulting in T-cell fatigue and immunological tolerance. This system is critical for limiting excessive immune responses that might harm normal tissues; nevertheless, it also permits tumors to avoid immune monitoring. Tumors can use this route to produce an immunosuppressive microenvironment, which protects them from an immune response and promotes cancer growth. PD-1 is an immunoglobulin superfamily member with structural similarities to CTLA-4, another immune checkpoint receptor. Unlike CTLA-4, which controls T-cell

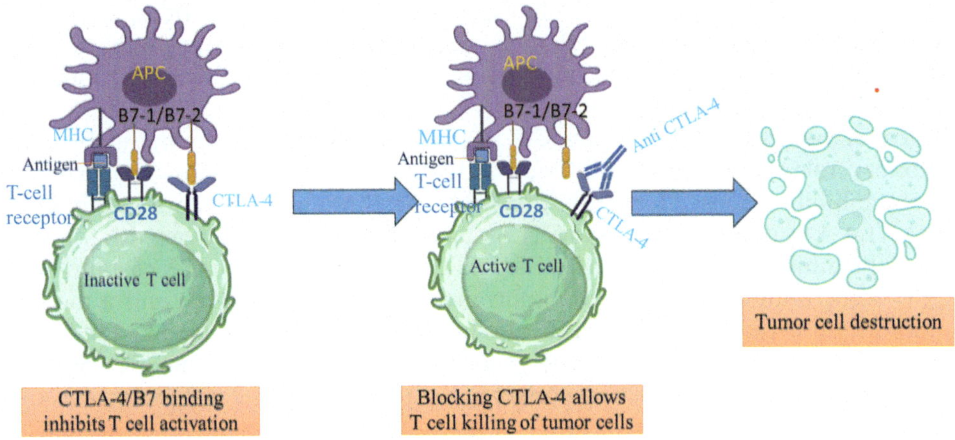

FIG. 3.2 CTLA-4 (cytotoxic T-lymphocyte-associated protein 4).

activation during the early phases of an immune response in lymphoid organs, PD-1 works mainly in peripheral tissues to inhibit T-cell activity during inflammatory reactions. This divergence underscores CTLA-4 and PD-1's complementary functions in immune response regulation. The PD-1/PD-L1 interaction recruits phosphatases such as SHP-2 to the immunological synapse, where they dephosphorylate key signaling molecules involved in TCR signaling, inhibiting downstream signaling pathways required for T-cell activation and proliferation (Deng et al., 2022).

Therapeutically targeting the PD-1/PD-L1 pathway entails inhibiting this interaction to restore T-cell activity and boost antitumor immunity. Several monoclonal antibodies have been designed to suppress PD-1 and PD-L1. Nivolumab and pembrolizumab block PD-1, whereas atezolizumab, durvalumab, and avelumab target PD-L1. These checkpoint inhibitors have demonstrated extraordinary success in treating a variety of malignancies, including melanoma, NSCLC, renal cell carcinoma, and Hodgkin's lymphoma. The therapeutic inhibition of PD-1 or PD-L1 reactivates fatigued T lymphocytes, restoring their capacity to recognize and destroy malignancy. This invigoration of T cells results in long-lasting responses and higher survival rates in many patients, cementing PD-1/PD-L1 inhibitors as a cornerstone of cancer immunotherapy.

Clinical trials have shown that PD-1/PD-L1 inhibitors provide considerable survival advantages, leading to their approval for a broad spectrum of cancers. Pembrolizumab, for example, has demonstrated promising outcomes in patients with advanced melanoma and NSCLC, outperforming standard treatment in terms of overall survival. Similarly, nivolumab has been successful in the treatment of metastatic melanoma, NSCLC, and renal cell carcinoma. These findings have led researchers to investigate PD-1/PD-L1 inhibitors in different cancer types further and in conjunction with other treatments to improve their efficacy (Bebars, 2023).

Despite their effectiveness, PD-1/PD-L1 inhibitors are not always successful, and a significant number of patients do not respond to them. This diversity in response underlines the need for predictive biomarkers to identify individuals who would benefit most from PD-1/

PD-L1 inhibition. Current research aims to find such biomarkers, which include PD-L1 expression levels on tumor cells, the presence of tumor-infiltrating lymphocytes, and the tumor's mutational load. Furthermore, understanding the causes of resistance to PD-1/PD-L1 inhibitors is critical for designing resistance-overcoming therapies that enhance patient outcomes.

Combination treatments using PD-1/PD-L1 inhibitors are being intensively researched to improve therapeutic effectiveness. Combining PD-1/PD-L1 inhibitors with other immune checkpoint inhibitors, such as CTLA-4 inhibitors, has shown potential in treating melanoma and renal cell carcinoma. These combinations are designed to target numerous inhibitory pathways, thereby boosting the antitumor immune response. Other combinations include PD-1/PD-L1 inhibitors combined with chemotherapy, radiation, targeted treatments, and other immunotherapies, such as cancer vaccines or oncolytic viruses. These techniques aim to enhance therapeutic outcomes by synergizing several mechanisms of action (Coman et al., 2024) (Fig. 3.3).

CAR-T cell therapy

CAR-T cell therapy is a breakthrough method of cancer treatment that uses the body's immune system to battle tumors. This therapy involves genetically modifying T lymphocytes to express chimeric antigen receptors (CARs), which target antigens on the surface of cancer cells. By altering these receptors, CAR-T cell therapy aims to instruct T cells to recognize and destroy cancer cells accurately, overcoming the limitations of standard cancer therapies and tumor immune evasion strategies.

CAR-T cell treatment begins with collecting T cells from the patient, a process often known as leukapheresis. Leukapheresis involves drawing blood from the patient, separating and collecting T cells, and returning the other blood components to the patient. This phase is critical because it provides the raw material—autologous T cells—from which CAR-T cells will be produced. Once the T cells have been harvested, they are sent to a specialized facility

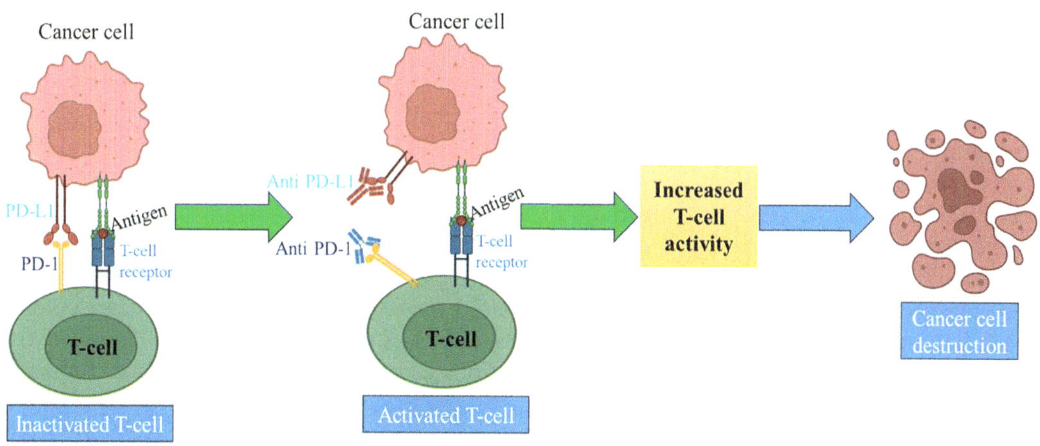

FIG. 3.3 PD-1/PD-L1 (programmed death-1/programmed death-ligand 1).

for genetic alteration. In the laboratory, T cells are genetically modified to express CARs on their surfaces (Huang et al., 2024).

CARs are synthetic receptors that combine an extracellular antigen recognition domain, often generated from a monoclonal antibody, with intracellular signaling regions that activate T cells in response to antigen binding. The CAR gene is injected into T cells by viral vectors, such as lentiviruses or retroviruses, which integrate the gene into the T-cell genome, ensuring long-term expression. This genetic modification technique converts regular T cells into potent CAR-T cells capable of attacking cancer cells that display the CAR-recognized antigen.

Following genetic alteration, CAR-T cells are grown in vitro to produce enough for therapeutic usage. During the expansion phase, genetically engineered T cells are cultured in the presence of cytokines such as interleukin-2 (IL-2), which promote proliferation and activation. The goal is to generate millions to billions of CAR-T cells from the initial sample, guaranteeing enough cells to mount an effective antitumor response when injected back into the patient. The CAR-T cells are continuously checked for viability, functionality, and purity throughout the procedure. Quality control mechanisms are in place to ensure that the CAR-T cells fulfill clinical requirements. CAR-T cells are collected, cleaned, and prepared for infusion after satisfactory growth.

Prior to the injection of CAR-T cells, patients often go through a conditioning phase, which includes chemotherapy. This conditioning serves many functions: it reduces tumor burden, improves the environment for CAR-T cell engraftment, and suppresses the patient's immune system to prevent rejection of the infused cells. Cyclophosphamide and fludarabine are two common medications used in conditioning regimens. After conditioning, the patient receives CAR-T cells via intravenous infusion. The injected CAR-T cells circulate through the bloodstream, searching for and attaching to cancer cells that express the target antigen. When the CAR binds, it activates T cells, causes them to proliferate, and induces cytotoxic activity, which kills cancer cells. This immune response is frequently followed by a rapid and significant reduction in tumor growth, demonstrating the decisive antitumor action of CAR-T cells (Wang et al., 2023).

CAR-T cell treatment has shown extraordinary success, notably in treating relapsed or refractory B-cell malignancies such as acute lymphoblastic leukemia and some kinds of non-Hodgkin lymphoma. Clinical investigations have revealed that many patients achieve complete remission, even after exhausting all other therapeutic options. The success of CAR-T cell therapy has resulted in the approval of numerous CAR-T cell products, such as tisagenlecleucel (Kymriah) and axicabtagene ciloleucel (Yescarta), which have transformed the therapeutic landscape for these challenging tumors. However, the therapy has its obstacles. Severe adverse effects, such as cytokine release syndrome (CRS) and neurotoxicity, can occur and must be carefully managed.

CRS is a systemic inflammatory response caused by CAR-T cells' rapid activation and multiplication, which produce significant amounts of cytokines. CRS symptoms can range from mild to life-threatening, necessitating the use of immunosuppressive medications such as tocilizumab, an IL-6 receptor antagonist. Neurotoxicity, also known as CAR-T cell-related encephalopathy syndrome, can cause disorientation, seizures, and cerebral edema, requiring close monitoring and intervention.

Another problem is antigen escape, which occurs when cancer cells lose or downregulate the target antigen, making it undetectable to CAR-T cells. This can lead to relapse and reduce

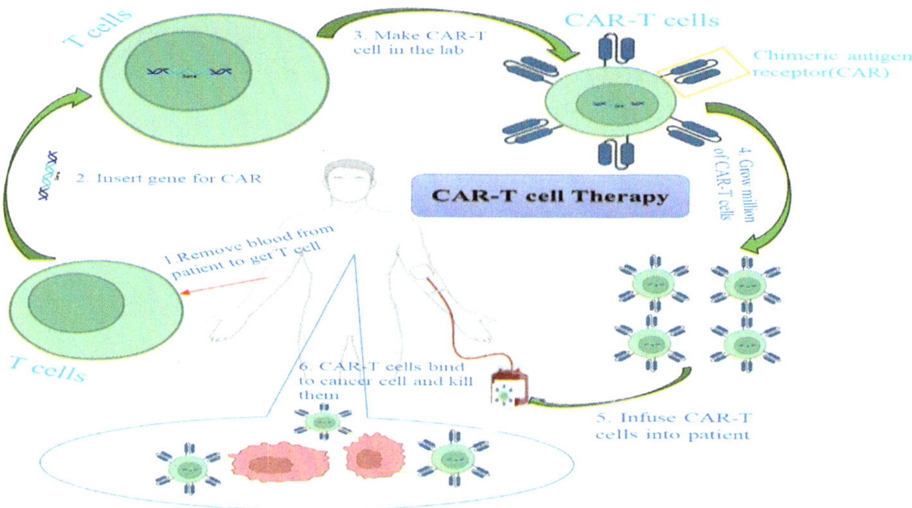

FIG. 3.4 Chimeric antigen receptor-T cell therapy.

the duration of responses. To address this, researchers are exploring techniques like dual-targeting CARs, which recognize numerous cancer cell antigens and reduce the possibility of antigen escape (Uscanga-Palomeque et al., 2023) (Fig. 3.4).

Cancer vaccines

Cancer vaccines are a new and promising cancer treatment method that uses the immune system to recognize and eliminate cancer cells. Cancer vaccines, as opposed to typical vaccinations for infectious illnesses, are intended to cure or prevent cancer recurrence by boosting the immune system to target TSAs. These vaccines overcome the immune system's tolerance to cancer cells, improving its capacity to identify and kill malignant cells (Zhang et al., 2024).

Types of cancer vaccines

Cancer vaccines are roughly categorized into preventative (prophylactic) and therapeutic vaccinations. Preventive vaccinations are designed to prevent cancer from developing in healthy people, generally by targeting viruses that cause cancer. For example, the human papillomavirus (HPV) and hepatitis B virus vaccines have been shown to be effective in preventing cervical and liver cancer, respectively. These vaccines generate an immune response to viral antigens, avoiding viral infections that can lead to cancer development.

Therapeutic vaccines, on the contrary, are intended to treat patients with cancer by generating an immune response against TSAs or TAAs. These vaccinations teach the immune system to recognize and kill cancer cells while sparing healthy cells. Therapeutic cancer vaccines are categorized into five kinds depending on their content and method of action:

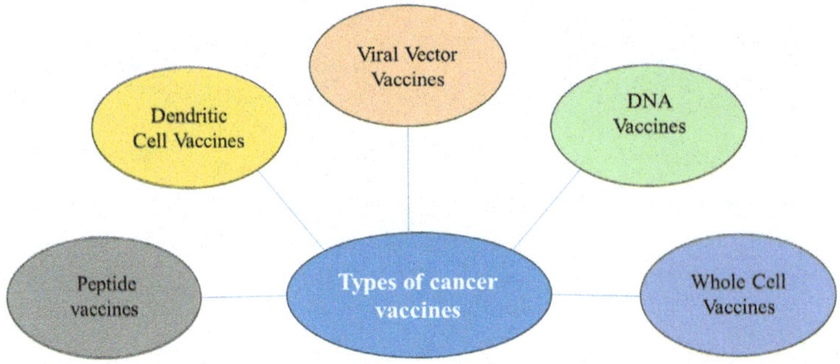

FIG. 3.5 Types of cancer vaccines.

peptide vaccines, dendritic cell (DC) vaccines, viral vector vaccines, DNA vaccines, and whole-cell vaccines (Fan et al., 2023) (Fig. 3.5).

Peptide vaccines

Peptide vaccines are a complex and tailored cancer immunotherapy method that activates the immune system to recognize and fight cancer cells. These vaccines comprise short sequences of amino acids called peptides that correspond to epitopes found in TAAs or TSAs. These epitopes are the regions of the antigen that T cells recognize, making them essential for eliciting a powerful and focused immune response. Peptide vaccines elicit a CTL response capable of precisely identifying and killing cancer cells that produce these antigens.

The development of peptide vaccines begins with the identification of relevant antigenic targets. This procedure frequently involves using modern genomic and proteomic tools to discover peptides expressed exclusively or predominantly by cancer cells. These peptides are then synthesized and combined into vaccinations that may be administered to patients. When the immune system encounters these peptides, it processes them and displays them on the surface of APCs using MHC molecules. This presentation is critical for activating T lymphocytes, particularly $CD8^+$ CTLs, which may destroy cancer cells presenting the target peptide (Zahedipour et al., 2023).

One of the primary benefits of peptide vaccinations is their specificity. Peptide vaccines, which target epitopes, have the potential to limit off-target effects and the danger of injuring normal, healthy tissues. This high level of specificity is instrumental in decreasing the side effects associated with more generalized cancer treatments such as chemotherapy and radiation. Furthermore, peptide vaccines may be engineered to incorporate numerous epitopes from distinct antigens, expanding the immune response and addressing tumor heterogeneity, which occurs when different cells within the same tumor express different antigens.

Despite their potential, peptide vaccines have faced several difficulties that have hampered their effectiveness. One fundamental difficulty is the intrinsic variety in immunological responses across individuals, which is determined by the genetic diversity of MHC molecules. Individuals have various MHC haplotypes. Therefore a peptide vaccination that works for one person may not work for another. This variety demands the development of peptide vaccines that can be recognized by a wide range of MHC molecules and vaccine personalization to fit individual patients' distinct MHC profiles.

Another fundamental difficulty is the immunosuppressive tumor microenvironment, which might reduce the function of vaccine-induced T cells. Tumors frequently use multiple techniques to avoid immune detection, including releasing immunosuppressive cytokines, recruiting Tregs, and expressing immunological checkpoint molecules such as PD-L1. These pathways can suppress the immune response, reducing the efficacy of peptide vaccines. To circumvent these hurdles, peptide vaccinations are frequently coupled with adjuvants, substances that boost the immune system. Adjuvants such as toll-like receptor (TLR) agonists, cytokines such as IL-2, and immune checkpoint inhibitors increase vaccination immunogenicity and counteract the tumor microenvironment's immunosuppressive effects (Yu et al., 2023).

Clinical studies using peptide vaccines have produced encouraging outcomes and emphasized the need for further optimization. Early research has shown that peptide vaccines can generate antigen-specific T-cell responses in individuals with various cancers, including melanoma, breast, and prostate. However, the translation of immune responses into therapeutic advantages, such as tumor reduction and increased survival, has been variable. This disparity highlights the complexities of the immune response and the necessity for a diverse strategy in cancer therapy.

Researchers are testing various novel approaches to improve the clinical effectiveness of peptide vaccines. One strategy is to deploy personalized neoantigen vaccinations customized to each individual's tumor's unique mutational landscape. Neoantigens are new peptides that result from tumor-specific mutations and are not found in normal tissues. Neoantigens are attractive targets for personalized vaccinations because they are specific to cancer cells. Advances in next-generation sequencing and analytics have enabled the fast identification and synthesis of neoantigen peptides, paving the way for highly personalized immunotherapies.

Another interesting approach is using multiepitope vaccines, which include peptides from numerous antigens or different sections of the same antigen. This method attempts to elicit a larger and more robust immunological response, limiting the possibility of immune escape and addressing tumor heterogeneity. Furthermore, peptide vaccines are being studied to combine with other immunotherapies, such as checkpoint inhibitors or adoptive T-cell transfer, to improve their therapeutic effectiveness. Checkpoint inhibitors, for example, can block the inhibitory signals that reduce T-cell activity, thereby increasing the efficacy of vaccine-induced immune responses (Buonaguro and Tagliamonte, 2023).

Advances in delivery technology aid the development of peptide vaccines as well. Nanoparticles, liposomes, and other delivery technologies are being investigated to enhance peptide vaccine stability, bioavailability, and targeted administration. These technologies can preserve peptides against degradation, increase their absorption by APCs, and ensure that they reach suitable immune compartments.

DC vaccines

DC vaccines are a critical and novel method in cancer immunotherapy that uses the body's immune cells to coordinate a targeted attack on cancer cells. DCs are potent APCs that help initiate and regulate immune responses. They excel in capturing, processing, and presenting antigens to T cells, acting as a link between the innate and adaptive immune systems. DC vaccines in cancer attempt to harness this natural potential to elicit a strong and targeted antitumor immune response.

The process of developing DC vaccines begins with the collection of mononuclear cells from the patient's blood using a method known as leukapheresis. These cells are subsequently grown in vitro and developed into DCs utilizing growth factors like granulocyte–macrophage colony-stimulating factor (GM-CSF) and IL-4. Once differentiated, DCs are loaded with tumor antigens. These antigens can come from various sources, such as tumor lysates, peptides, proteins, RNA, and even complete tumor cells. The selection of the antigen source is critical since it dictates the scope and specificity of the immune response (Macri et al., 2023).

Tumor lysates and entire tumor cells contain a wide range of antigens, which can help target heterogeneous tumors. In contrast, peptide or protein antigens can be chosen for their specificity, targeting unique or overexpressed proteins in cancer cells.

After being loaded with antigens, DCs mature. Mature DCs express significant quantities of MHC molecules, costimulatory molecules (such as CD80 and CD86), and cytokines, all required for successful T-cell activation. These mature, antigen-loaded DCs are injected into the patient via intradermal or subcutaneous routes. Following injection, DCs move to lymph nodes and present tumor antigens to T cells, namely, CTLs. This presentation stimulates T lymphocytes, which multiply and develop into effector cells capable of recognizing and eliminating cancer cells that express the same antigens.

One of the primary benefits of DC vaccines is their capacity to elicit a diverse immune response. DC vaccines can initiate a coordinated attack on cancer cells by delivering antigens to $CD8^+$ CTLs and $CD4^+$ helper T cells. This results in direct cytotoxicity and helper T-cell support, which enhances and sustains the CTL response. Furthermore, DCs can train the immune system to recognize and respond to cancer antigens precisely, reducing the chance of injury to healthy tissues and minimizing side effects.

Clinical trials with DC vaccines have yielded encouraging outcomes, particularly in malignancies resistant to conventional treatments. Sipuleucel-T (Provenge) is the first FDA-approved DC vaccine to treat metastatic castration-resistant prostate cancer. It improved survival rates in clinical studies, showing the promise of DC-based immunotherapy. In this vaccine, patient-derived DCs are loaded with a fusion protein of prostatic acid phosphatase, an antigen found in prostate cancer cells, and GM-CSF, which enhances DC maturation. The modified DCs are reintroduced into the patient, stimulating T lymphocytes that target and destroy prostate cancer cells (Lee et al., 2023).

Despite the success of Sipuleucel-T, the widespread use of DC vaccines confronts significant hurdles. The complexity and expense of the production process present a substantial challenge. DC vaccine manufacture consists of several phases, including leukapheresis, cell culture, antigen loading, maturation, and quality control, all of which must be customized to each patient. While highly specialized, this personalized technique is labor-intensive and costly, restricting its adoption. Additionally, the immunosuppressive tumor microenvironment might reduce the efficacy of DC vaccines. Tumors can evade immune detection by expressing immune checkpoint molecules (e.g., PD-L1), secreting immunosuppressive cytokines (e.g., TGF-β), and recruiting Tregs and MDSCs. These factors can diminish the immunological response induced by DC vaccines, limiting their therapeutic effectiveness.

To address these problems, researchers are investigating various techniques to improve the efficacy of DC vaccines. One method is to combine DC vaccines with immune checkpoint drugs. These inhibitors, such as PD-1/PD-L1 or CTLA-4 blockers, can relax the immune system, allowing T cells to launch a more powerful onslaught on cancer cells. Combining DC vaccines with other immunomodulatory drugs, such as cytokines (e.g., IL-2, IL-12) or adjuvants (e.g., TLR agonists), might improve DC maturation and activation, thereby increasing the overall immune response (Qian et al., 2023).

Another interesting approach is the use of next-generation DC vaccines. Genetic engineering and synthetic biology advances have made it possible to create DCs that express several costimulatory molecules, release immunostimulatory cytokines, or withstand the suppressive effects of the tumor microenvironment. For example, genetically modified DCs that produce constitutively active versions of essential signaling molecules can boost T-cell activation and antitumor responses. Furthermore, introducing nanoparticle-based delivery methods can increase the efficiency of antigen loading and presentation, making the manufacturing process more scalable and cost-efficient.

Viral vector vaccines

Viral vector vaccines are a cutting-edge method in cancer immunotherapy that leverages the unique capabilities of viruses to transfer genetic material encoding TSAs into the body's cells. These vaccines employ modified viruses as vectors, designed to be safe and nonreplicating, to insert genes that direct cells to create antigens, which cause a strong immune response against cancer cells. This procedure combines the benefits of both vaccination and gene therapy, resulting in a highly focused and powerful strategy for evoking antitumor immunity.

The process of viral vector vaccines begins with the choice of a suitable viral vector. Adenoviruses, lentiviruses, vaccinia, and adeno-associated viruses (AAV) are among the most often utilized vectors. These vectors are chosen because of their capacity to infect human cells effectively and their well-documented safety characteristics. The viral vectors are genetically modified to contain genes that encode tumor antigens or immunostimulatory molecules. Importantly, these vectors are designed to be replication-deficient, allowing them to infect cells and deliver their genetic payload without reproducing themselves, ensuring safety.

Once conceived and manufactured, the viral vector is supplied to the patient via intramuscular, subcutaneous, or intravenous injection. Following delivery, the viral vector infects various cells, including DCs and other APCs. The vector's genetic material reaches the host cell's nucleus, where it is transcribed and translated into the encoded tumor antigens. The host cells subsequently digest these antigens and display them on their surface, attached to MHC molecules (Lundstrom, 2023).

The presentation of these antigens by APCs is critical for activating the adaptive immune system. DCs move to lymph nodes and engage with T cells after infecting the viral vector and displaying tumor antigens. This antigen presentation is combined with costimulatory signals produced by APCs, which are required for complete T-cell activation. $CD8^+$ CTLs recognize the antigen–MHC class I complex, which activates, proliferates, and differentiates into effector T cells capable of directly destroying tumor cells. Concurrently, $CD4^+$ helper T cells recognize the antigen–MHC class II complex, which facilitates an enormous immune response, such as B-cell activation and antibody production.

Viral vector vaccines provide various benefits over regular vaccination methods. One notable benefit is that they can stimulate both cellular and humoral immunity. The robust T-cell responses induced by viral vectors are especially crucial for targeting tumors, which are frequently resistant to antibody-mediated elimination. Furthermore, the use of viral vectors enables the inclusion of many antigens or immunomodulatory genes into a single vaccination, increasing the breadth and effectiveness of the immune response. This multi-antigen method is beneficial for tackling tumor heterogeneity and minimizing the possibility of immune escape, which occurs when cancer cells evolve and elude immune system identification (Busen et al., 2024).

Clinical experiments with viral vector vaccines have yielded encouraging outcomes in various malignancies. For example, PROSTVAC, a viral vector vaccine based on vaccinia and fowlpox viruses, encodes prostate-specific antigen (PSA) and costimulatory components to boost immune response. In clinical studies, PROSTVAC was shown to produce robust PSA-specific T-cell responses and increase survival in patients with metastatic prostate cancer. Similarly, Talimogene laherparepvec (T-VEC), an oncolytic herpes simplex virus type 1 modified to express GM-CSF, is effective in treating melanoma by directly lysing tumor cells and activating systemic antitumor immunity (Busen et al., 2024).

DNA vaccines

DNA vaccines are a novel and promising cancer immunotherapy method that uses the body's cellular machinery to create tumor antigens and induce a robust immune response. Unlike conventional vaccinations, which often contain inactivated pathogens or recombinant proteins, DNA vaccines use genetically designed plasmids to encode the desired antigens. This approach has various advantages, including simplicity of manufacture, stability, and the capacity to generate both humoral and cellular immune responses.

The discovery of TSAs is the first step in developing a DNA vaccination. These antigens are often proteins expressed exclusively or primarily by cancer cells. Once identified, the genes that encode these antigens are introduced into a plasmid vector. Plasmids are circular DNA molecules that can multiply without chromosomal DNA, making them excellent vectors for genetic material delivery. The plasmid also contains a powerful promoter, ensuring high antigen production inside the host cells (Blakney and Bekker, 2022).

When administered intramuscularly or intradermally, the plasmid DNA is taken up by host cells at the injection site, including muscle cells and DCs. Electroporation, which uses brief electrical pulses to promote cell membrane permeability, is commonly used to improve DNA uptake. Once inside the cell, the plasmid DNA goes to the nucleus, where the host cell's transcriptional machinery converts it into mRNA. This mRNA is subsequently translated into tumor antigen proteins in the cytoplasm.

The newly synthesized tumor antigens are processed within the cell. The proteasome converts endogenous antigens into peptide fragments, which are then loaded onto MHC class I molecules. These MHC class I-antigen complexes are delivered to the cell's surface and recognized by $CD8^+$ CTLs. This identification is necessary to activate and multiply CTLs, the primary effector cells in the antitumor immune response. Furthermore, some generated antigens are secreted or released after cell lysis and are picked up by APCs, such as DCs. These APCs process antigens and deliver them to $CD4^+$ helper T cells via MHC class II molecules, boosting the immune response by enhancing CTL activation and B-cell-mediated antibody production (Pagliari et al., 2023).

DNA vaccines' primary benefit is their capacity to elicit a comprehensive and long-lasting immune response. These vaccines ensure continual antigen generation by directly encoding tumor antigens into DNA, resulting in sustained immune activation. This contrasts with protein-based vaccinations, where the body swiftly destroys and eliminates the antigen. DNA vaccines offer a good safety profile because they do not include live pathogens or replication-competent viruses, which reduces the danger of infection or unintentional genomic integration.

The simplicity and durability of DNA vaccines make them ideal for clinical application. Plasmid DNA is straightforward and affordable to create in vast numbers, and it remains stable at room temperature, facilitating storage and distribution. This is especially useful in resource-constrained environments, where cold chain logistics are complex.

Preclinical research has shown that DNA vaccines can be effective in various cancer types. This research has demonstrated that DNA vaccines may elicit robust CTL responses in animal models, resulting in considerable tumor reduction and longer life. Clinical experiments in people have also produced encouraging outcomes. For example, VGX-3100, a DNA vaccine that targets the HPV16 and HPV18 E6 and E7 oncogenes, has been demonstrated to be effective in treating high-grade cervical intraepithelial neoplasia, a precancerous disease caused by HPV infections. The vaccination elicited significant T-cell responses, resulting in the reduction of lesions and viral clearance.

Despite their potential, DNA vaccines confront various hurdles that must be overcome to ultimately realize their therapeutic potential. One fundamental difficulty is naked DNA's very poor immunogenicity compared to other vaccination platforms. Enhancing the delivery and expression of DNA vaccines is an essential field of study. Electroporation has been proven to dramatically enhance DNA absorption and antigen expression; however, it requires specialized equipment and may induce pain at the injection site (Pagliari et al., 2023).

To address these restrictions, researchers are experimenting with various approaches. One approach is to employ molecular adjuvants—genes that encode immunostimulatory molecules supplied alongside the plasmid that encodes the antigens. These adjuvants can enhance DC activation, T-cell priming, and the overall immunological response. Examples include genes that produce cytokines (e.g., IL-2, IL-12), costimulatory molecules (e.g., CD80, CD86), and TLR ligands.

Another exciting technique is to employ nanoparticle-based delivery methods. Encapsulating plasmid DNA in nanoparticles can preserve it from degradation, boost cellular absorption, and increase the overall efficacy of antigen expression. Nanoparticles can also target specific cell types, improving the accuracy and efficacy of DNA vaccinations (Pagliari et al., 2023).

Whole-cell vaccines

Whole-cell vaccines are a potential cancer immunotherapy method that uses whole tumor cells, either autologous (from the patient) or allogeneic (from other people with the same disease), to trigger an immune response. This technique utilizes the entire repertoire of tumor antigens found in cancer cells, resulting in a comprehensive antigenic profile that can overcome tumor heterogeneity and antigenic variation. Whole-cell vaccines can elicit a strong and broad immune response, targeting numerous tumor antigens simultaneously and reducing the probability of immune escape.

Collecting tumor cells is the first step in making whole-cell vaccines. Autologous vaccines are created by harvesting tumor cells directly from the patient's tumor, guaranteeing that the antigens are unique to their malignancy. Tumor cells for allogeneic vaccinations are obtained from existing cell lines or other people suffering from the same malignancy, resulting in a standardized and scalable technique. Once collected, these cells are usually inactivated so they cannot reproduce or cause illness. This inactivation is frequently accomplished by irradiation, chemical treatment, or heating (Keenan and Jaffee, 2012).

Inactivated tumor cells are then employed as the vaccination. APCs, such as DCs and macrophages, take up these cells after delivery, often via intradermal or subcutaneous injection. The APCs break down the tumor cells into smaller peptide fragments. These fragments are subsequently loaded onto MHC molecules and placed on the surface of APCs. MHC class I molecules present antigens to $CD8^+$ CTLs, whereas MHC class II molecules present antigens to $CD4^+$ helper T cells.

The exposure of these antigens stimulates the adaptive immune system. $CD8^+$ CTLs recognize tumor antigens presented by MHC class I molecules and become activated, proliferate, and differentiate into effector T cells. These effector T cells can then seek out and destroy tumor cells that express the same antigens. Concurrently, $CD4^+$ helper T lymphocytes recognize antigens presented by MHC class II molecules and facilitate CTL activation and proliferation. They also aid in activating B cells, which generate antibodies that target tumor antigens and enhance the overall immune response.

Whole-cell vaccines have numerous substantial benefits. One of the key advantages is the extensive antigenic coverage they offer. Unlike vaccinations targeting a single antigen, whole-cell vaccines provide a diverse set of known and undiscovered tumor antigens. This comprehensive antigenic profile is beneficial for tackling tumor heterogeneity, in which different cells within the same tumor or between tumors in the same patient may express distinct antigens. Whole-cell vaccines can produce a more robust and flexible immune response by exposing a wide range of antigens, limiting the risk of immunological escape due to any antigenic variant (Keenan and Jaffee, 2012).

Another benefit of whole-cell vaccines is their capacity to stimulate cellular and humoral immune responses. CTL activation results in direct tumor cell death, whereas helper T- and B-cell activation results in antibody formation, which can assist in targeting and eliminating tumor cells. This dual-action method improves the immune response's overall efficacy against malignancy.

Clinical studies of whole-cell vaccinations have yielded promising outcomes in various cancers. For example, granulocyte--macrophage colony-stimulating factor-expressing vaccine (GVAX), an allogeneic whole-cell vaccine, has been intensively researched in pancreatic and other cancers. GVAX comprises tumor cells that have been genetically engineered to release GM-CSF, a cytokine that stimulates DC activation and maturation. Clinical trials have shown that, when paired with immune checkpoint inhibitors, GVAX can stimulate significant immune responses and enhance survival in patients with advanced pancreatic cancer.

Despite their potential, whole-cell vaccines face various hurdles that must be overcome to maximize therapeutic effectiveness. One of the major problems is the possibility of limited immunogenicity. Tumor cells, especially in the setting of autologous vaccinations, may not be immunogenic enough to trigger a robust immune response on their own. This might be due to tumor-induced immunosuppression or a lack of essential costimulatory signals. Various approaches are being investigated to improve the immunogenicity of whole-cell vaccines. One method is to utilize genetic alteration to express immunostimulatory molecules. For example, altering tumor cells to generate GM-CSF can enhance DC recruitment and activation, thereby improving the immune response. Another technique is combining whole-cell vaccines with adjuvants, chemicals that stimulate the immune system's reaction to an antigen. TLR agonists, cytokines, and other immunomodulatory drugs can dramatically increase the efficacy of whole-cell vaccines (Keenan and Jaffee, 2012).

Another issue is the scalability and standardization of vaccine manufacturing. While autologous whole-cell vaccines are personalized and unique to each patient, their manufacturing is complicated, time-consuming, and costly. On the contrary, allogeneic whole-cell vaccines provide a more standardized and scalable method, although they may need careful matching to ensure compatibility and prevent immune rejection. Advances in cell culture methods, biomanufacturing, and genetic engineering are contributing to resolving these difficulties, making whole-cell vaccine production more viable and cost-effective.

Innovative methods are also being investigated to enhance the targeting and distribution of whole-cell vaccines. Combining whole-cell vaccines with immune checkpoint inhibitors, such as anti-PD-1 or anti-CTLA-4 antibodies, can boost the immune response by counteracting tumors' immunosuppressive processes. Furthermore, nanoparticles, liposomes, and other delivery vehicles can increase whole-cell vaccines' stability, targeting, and absorption, thereby increasing their therapeutic potential (Szwejser-Zawislak et al., 2023).

Conclusion

The immune system's dual involvement in cancer, which protects against tumor development while also aiding tumor progression when evasion occurs, highlights the disease's complexity. Advances in understanding these systems have resulted in immunotherapies, such as immune checkpoint inhibitors and cancer vaccines, which aim to use the immune system to combat cancer more effectively. CAR-T cell therapy has demonstrated the ability of genetically modified T cells to target and destroy cancer cells with remarkable precision. In contrast, cancer vaccines, such as peptides, DCs, viral vectors, DNA, and whole-cell vaccines, have opened new avenues for personalized and preventive cancer treatment. The use of oncolytic viruses demonstrates the new ways being explored to eliminate tumor cells and promote systemic immune responses to cancer. Despite the obstacles, such as immune evasion by tumors and the complexities of administering effective therapies, ongoing advances in understanding and modulating the immune system's involvement with cancer provide significant hope for improving patient outcomes.

References

Azimnasab-sorkhabi, P., Soltani-asl, M., Kfoury Junior, J.R., 2023. Cytotoxic T-lymphocyte-associated protein 4 (CTLA-4) as an undetermined tool in tumor cells. Hum. Cell. 1749-0774. 36 (4), 1225–1232. Available from: https://www.springer.com/journal/13577. doi:10.1007/s13577-023-00893-8.

Bebars, S.M., 2023. Immune checkpoint inhibitors programmed cell death-1/programmed cell death-ligand1 (PD-1/PD-L1) for cancer therapy. In: Immune Checkpoint Inhibitors-New Insights and Recent Progress. IntechOpen, pp. 305-319.

Blakney, A.K., Bekker, L.G., 2022. DNA vaccines join the fight against COVID-19. Lancet. 1474-547X. 399 (10332), 1281–1282. Available from: http://www.journals.elsevier.com/the-lancet/. doi:10.1016/S0140-6736(22)00524-4.

Buonaguro, L., Tagliamonte, M., 2023. Peptide-based vaccine for cancer therapies. Front. Immunol. 1664-3224. 14. doi:10.3389/fimmu.2023.1210044.

Busen, W., Jinghan, X., Zhiqiang, G., 2024. Advances in virus-vectored vaccines. Synth. Biol. J. 5, 281.

Coman, O., Grigorescu, B.-L., Huţanu, A., Bacârea, A., Văsieşiu, A.M., Fodor, R.Ş., Stoica, F., Azamfirei, L., 2024. The role of programmed cell death 1/programmed death ligand 1 (PD-1/PD-L1) axis in sepsis-induced apoptosis. Medicina. 1648-9144. 60 (7), 1174. doi:10.3390/medicina60071174.

Deng, J., Cheng, Z., Long, J., Dömling, A., Tortorella, M., Wang, Y., 2022. Small molecule inhibitors of programmed cell death ligand 1 (PD-L1): a patent review (2019–2021). Expert Opin. Ther. Pat. 1354-3776. 32 (5), 575–589. doi:10.1080/13543776.2022.2045276.

Di Giacomo, A.M., Lahn, M., Eggermont, A.M.M., Fox, B., Ibrahim, R., Sharma, P., Allison, J.P., Maio, M., 2024. The future of targeting cytotoxic T-lymphocyte-associated protein-4: is there a role? Eur. J. Cancer. 0959-8049. 198, 113501. doi:10.1016/j.ejca.2023.113501.

Emens, L.A., Loi, S., 2023. Immunotherapy approaches for breast cancer patients in 2023. Cold Spring Harb. Perspect. Med. 2157-1422. 13 (4), a041332. doi:10.1101/cshperspect.a041332.

Fan, T., Zhang, M., Yang, J., Zhu, Z., Cao, W., Dong, C., 2023. Therapeutic cancer vaccines: advancements, challenges and prospects. Signal Transduct. Target. Ther. 2059-3635. 8 (1). doi:10.1038/s41392-023-01674-3.

Huang, Q., Li, H., Zhang, Y., 2024. A bibliometric and knowledge-map study on the treatment of hematological malignancies with CAR-T cells from 2012 to 2023. Hum. Vaccin. Immunother. 2164-5515. 20 (1). doi:10.1080/21645515.2024.2371664.

Huang, X., Guo, J., Wang, S., 2012. Global research trends on B7-H3 for cancer immunotherapy: a bibliometric analysis. Hum. Vaccin. Immunother. 19, 2246498.

Iglesias-Escudero, M., Arias-González, N., Martínez-Cáceres, E., 2023. Regulatory cells and the effect of cancer immunotherapy. Mol. Cancer. 1476-4598. 22 (1). doi:10.1186/s12943-023-01714-0.

Keenan, B.P., Jaffee, E.M., 2012. Whole cell vaccines—past progress and future strategies. Semin. Oncol. 0093-7754. 39 (3), 276–286. doi:10.1053/j.seminoncol.2012.02.007.

Lee, K.W., Yam, J.W.P., Mao, X., 2023. Dendritic cell vaccines: a shift from conventional approach to new generations. Cells. 2073-4409. 12 (17). Available from: https://www.mdpi.com/journal/cells. doi:10.3390/cells12172147.

Li, B., Guo, J.J., 2023. Immune checkpoint inhibitors combined with targeted therapy: the recent advances and future potentials. Cancers 15, 2858.

Li, Y.-N., Xie, B., Zhang, Y., 1999. Advances and key focus areas in gastric cancer immunotherapy: a comprehensive scientometric and clinical trial review. World J. Gastroenterol. 29, 5593.

Lundstrom, K., 2023. Viral vectors in gene therapy: where do we stand in 2023? Viruses. 1999-4915. 15 (3), 698. doi:10.3390/v15030698.

Macri, C., Jenika, D., Ouslinis, C., Mintern, J.D., 2023. Targeting dendritic cells to advance cross-presentation and vaccination outcomes. Semin. Immunol. 1044-5323. 68, 101762. doi:10.1016/j.smim.2023.101762.

Obeagu, E., Obeagu, G., 2024. Cytotoxic T-lymphocyte-associated protein 4 (CTLA-4) blockade and HIV-associated kaposi sarcoma: a promising therapeutic strategy. Elite J. Sci. Res. Rev. 2, 78–94.

Pagliari, S., Dema, B., Sanchez-Martinez, A., Montalvo Zurbia-Flores, G., Rollier, C.S., 2023. DNA vaccines: history, molecular mechanisms and future perspectives. J. Mol. Biol. 0022-2836. 435 (23), 168297. doi:10.1016/j.jmb.2023.168297.

Qian, D., Li, J., Huang, M., Cui, Q., Liu, X., Sun, K., 2023. Dendritic cell vaccines in breast cancer: immune modulation and immunotherapy. Biomed. Pharmacother. 0753-3322. 162, 114685. doi:10.1016/j.biopha.2023.114685.

Szwejser-Zawislak, E., Wilk, M.M., Piszczek, P., Krawczyk, J., Wilczyńska, D., Hozbor, D., 2023. Evaluation of whole-cell and acellular pertussis vaccines in the context of long-term herd immunity. Vaccines. 2076-393X. 11 (1). Available from: http://www.mdpi.com/journal/vaccines. doi:10.3390/vaccines11010001.

Tay, C., Tanaka, A., Sakaguchi, S., 2023. Tumor-infiltrating regulatory T cells as targets of cancer immunotherapy. Cancer Cell. 1878-3686. 41 (3), 450–465. Available from: https://www.journals.elsevier.com/cancer-cell. doi:10.1016/j.ccell.2023.02.014.

Uscanga-Palomeque, A.C., Chávez-Escamilla, A.K., Alvizo-Báez, C.A., Saavedra-Alonso, S., Terrazas-Armendáriz, L.D., Tamez-Guerra, R.S., Rodríguez-Padilla, C., Alcocer-González, J.M., 2023. CAR-T cell therapy: from the shop to cancer therapy. Int. J. Mol. Sci. 1422-0067. 24 (21). Available from: http://www.mdpi.com/journal/ijms. doi:10.3390/ijms242115688.

Wang, V., Gauthier, M., Decot, V., Reppel, L., Bensoussan, D., 2023. Systematic review on CAR-T cell clinical trials up to 2022: academic center input. Cancers. 2072-6694. 15 (4), 1003. doi:10.3390/cancers15041003.

Yu, X., Mai, Y., Wei, Y., Yu, N., Gao, T., Yang, J., 2023. Therapeutic potential of tolerance-based peptide vaccines in autoimmune diseases. Int. Immunopharmacol. 1567-5769. 116, 109740. doi:10.1016/j.intimp.2023.109740.

Zahedipour, F., Jamialahmadi, K., Zamani, P., Reza Jaafari, M., 2023. Improving the efficacy of peptide vaccines in cancer immunotherapy. Int. Immunopharmacol. 1567-5769. 123, 110721. doi:10.1016/j.intimp.2023.110721.

Zhang, T., Zhang, Y., Wei, J., 2024. Overcoming the challenges encountered in CAR-T therapy: latest updates from the 2023 ASH annual conference. Front. Immunol. 1664-3224. 15. doi:10.3389/fimmu.2024.1389324.

CHAPTER

4

Combination therapies: Synergistic approaches in cancer treatment

Indrani Devi Sarma
Department of Pharmacology, AIIMS, Guwahati, India

Introduction

Cancer treatment has evolved significantly over the years, with researchers constantly seeking new and improved therapies to combat this complex disease. Combination therapy, which uses multiple medications or treatment methods at the same time to target several elements of tumor growth and spread, is one potential strategy that has gained popularity recently. By combining treatments with complementary mechanisms of action, healthcare providers hope to achieve greater efficacy and reduce the likelihood of the development of drug resistance. However, the effectiveness of combination therapy in treating cancer depends on both a careful selection of the right medications and dosage schedules and a deep comprehension of the disease's underlying causes. In addition, to guarantee the safety and well-being of patients receiving combination therapy, attentive monitoring and management of possible side effects and drug interactions are essential.

Background on cancer treatment

With the advent of novel treatments such as serplulimab and abemaciclib, the field of cancer treatment has seen tremendous change throughout time, and these treatments have shown promise in the treatment of various cancer types. Serplulimab, a novel treatment for extensive-stage small cell lung cancer, has demonstrated encouraging outcomes that include an increase in quality-adjusted life years and improved overall survival when combined with chemotherapy. Abemaciclib, a cyclin-dependent kinase 4/6 inhibitor, is used as an adjuvant treatment for hormone receptor-positive, HER2 (Human Epidermal Growth Factor Receptor 2) negative early breast cancer. Despite the efficacy of these therapies, the financial implications and costeffectiveness considerations play a crucial role in decision-making for both patients and healthcare providers when opting for combination therapies in cancer treatment.

Pharmacological Approaches in Cancer Treatment
https://doi.org/10.1016/B978-0-443-38307-6.00002-8

The rationale for combination therapy in cancer

Enhanced efficacy

Using a combination of different drugs or therapies can lead to a more effective treatment than relying on a single agent. Each drug targets distinct pathways, creating a synergistic or additive effect that can be more powerful than any individual treatment alone.

Overcoming resistance

Cancer cells can develop resistance to individual treatments. Combining multiple agents can help prevent or delay the development of resistance, ensuring that the treatment remains effective for a longer duration.

Reducing side effects

By combining drugs that work through different mechanisms, the overall toxicity of the treatment can be minimized. This approach is particularly useful for patients who suffer from severe side effects with single-agent therapies (Fig. 4.1).

The rationale for combination therapy in cancer lies in its ability to target multiple pathways simultaneously, thereby increasing the likelihood of successful treatment outcomes. Physicians can prevent the emergence of drug resistance and improve treatment efficacy by combining different types of drugs with distinct mechanisms of action. This method also permits the use of lower dosages of individual medications, reducing the possibility of harmful side effects while preserving therapeutic effectiveness. In addition, combining

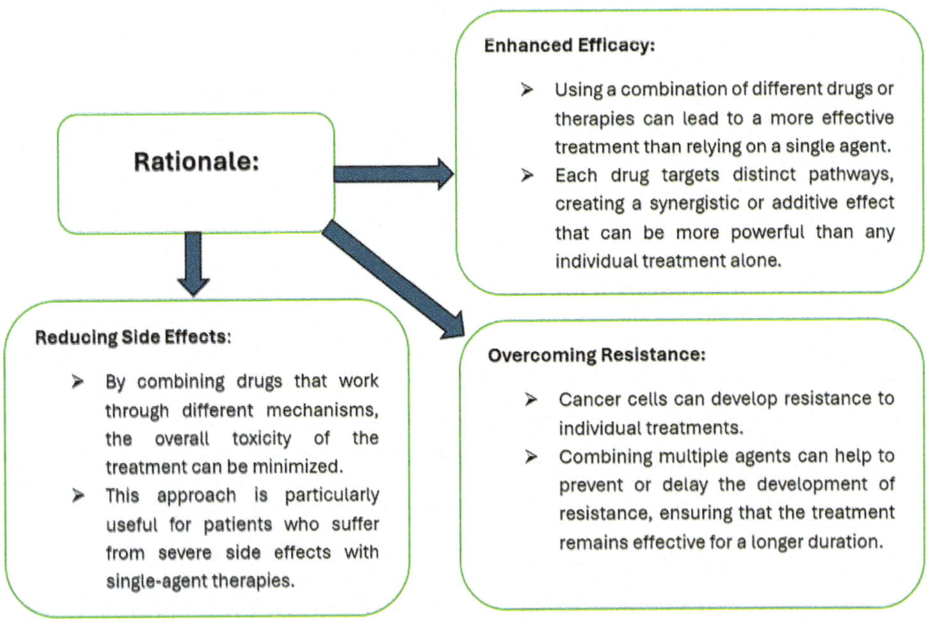

FIG. 4.1 Rationale for combination therapy. Three aspects in the rationale of combination therapies in cancer: enhancing efficacy, overcoming resistance, and reducing side effects.

targeted therapies with traditional chemotherapy agents can harness its synergistic effects, leading to increased tumor cell death and improved patient survival rates. Studies have shown that combination therapy has the potential to overcome the limitations of single-agent treatments, offering a promising strategy for combating the heterogeneity and adaptability of cancer cells. Overall, the use of combination therapy represents a rational and effective approach to the fight against cancer.

Definition

Combination therapy in cancer treatment involves the strategic use of multiple drugs to target various aspects of tumor growth and development, aiming to enhance efficacy and overcome drug resistance. Research involving the exploration of metastasis-directed therapy for recurrent prostate cancer highlights the potential benefits of combining localized treatment with systemic approaches. This approach signifies a shift toward multimodal strategies that may improve patient outcomes. In addition, other studies have introduced the concept of sequential drug administration as a promising alternative to simultaneous drug delivery, emphasizing the importance of precise timing and switching strategies to enhance therapeutic effects. By considering these findings, physicians can better understand the complexities and significance of combination therapy in oncology, paving the way for more tailored and effective treatment regimens.

Recent advancements in the field of cancer therapy have highlighted the potential benefits of implementing combination therapy in clinical practice. The analysis of circulating tumor cells (CTCs), as well as CTC-related transcripts in small cell lung cancer (SCLC) patients, has shown promising results in predicting outcomes and guiding treatment decisions. The discovery of neuroendocrine and epithelial transcripts and the association between high CTC counts and overall survival demonstrate the value of liquid biopsy analysis in SCLC. Furthermore, in a case study involving colistin medication, the use of lipidomics in therapeutic drug monitoring has demonstrated potential in optimizing treatment regimens. This combined approach not only helps to confirm effective drug dosing but also provides insights into lipid changes that could aid in managing antimicrobial and inflammatory responses in critically ill patients. Therefore the implementation of combination therapy incorporating novel biomarkers and analytical techniques holds promise for improving clinical outcomes in cancer and other diseases.

Types of combination therapies

Combining immunotherapy and targeted therapy is another kind of combination therapy used to treat cancer. Targeted therapy concentrates on particular molecular targets that are important in the development and growth of cancer cells. By combining this targeted therapy with immunotherapy, healthcare providers can potentially enhance the immune response against cancer cells while also directly targeting the molecular pathways driving tumor growth. This strategy has demonstrated encouraging outcomes in a variety of cancer types, including melanoma and several forms of lung cancer. In addition, hormone therapy, when combined with chemotherapy, has been studied as a possible treatment for breast cancer that is hormone receptor-positive. Previous research (Akhade et al., 2022; Arruebo et al., 2011;

Cattrini et al., 2018; DeForge et al., 2023; Delou et al., 2019; Doostmohammadi et al., 2024; Ezike et al., 2023; Garg et al., 2024; Gerhardtova et al., 2024; Hayashi et al., 2013; Huo et al., 2023; Liu et al., 2024; MU et al., 2014; Mokhtari et al., 2017; Mondelo-Macía et al., 2021; Nitiss, 2009; Peng et al., 2023; Peters and Brown, 2015; Tolcher et al., 2023; Vanneman and Dranoff, 2012; Welsch et al., 2024; Zheng et al., 2024) has shown that this combination approach can lead to improved outcomes in rates of survival and disease progression compared to using either therapy alone. Furthermore, combining different types of chemotherapy agents can also be effective in treating certain types of cancer. By using drugs that target different mechanisms of action, such as DNA replication or cell division, the effectiveness of treatment can potentially be maximized while minimizing the risk of drug resistance.

Clinically used synergistic strategies

1. **Surgery, radiation, and chemotherapy**: Combining radiation therapy, chemotherapy, and other cancer medications with surgery is an optimal strategy for several forms of cancer.
2. **Combining radio/chemotherapy with immunotherapy**: This approach combines radiotherapy or chemotherapy with immunotherapy. It has shown promise in treating various cancer types, including lung and gastric cancers.

The advent of antibodydrug conjugates

Researchers are actively investigating novel combination therapies aimed at improving cancer treatment efficacy. These approaches involve combining different classes of drugs, such as antibodydrug conjugates (ADCs) with chemotherapy, molecularly targeted agents, and immunotherapy. While promising, these experimental strategies require rigorous evaluation to ensure safety and effectiveness.

Definition

ADCs combine the ability of cytotoxic drugs to kill cancer with the targeting capabilities of monoclonal antibodies (mAbs). ADCs target only tumor cells while sparing healthy cells.

Parts of ADC

An ADC consists of the following three components (Fig. 4.2): **antigen binding site**, where the mAb specifically recognizes a tumor antigen (protein) found on cancer cells; **linker**, which connects the antibody to the cytotoxic payload; and **cytotoxic payload**, the actual anticancer drug. The antibody binds to tumor cell surface antigens, triggering internalization of the ADC. Once inside the cancer cell, the cytotoxic payload is released, leading to cell death.

Examples

1. **Trastuzumab emtansine (T-DM1)**: Approved for metastatic breast cancer that is HER2-positive. It combines a cytotoxic drug with trastuzumab, an anti-HER2 antibody.

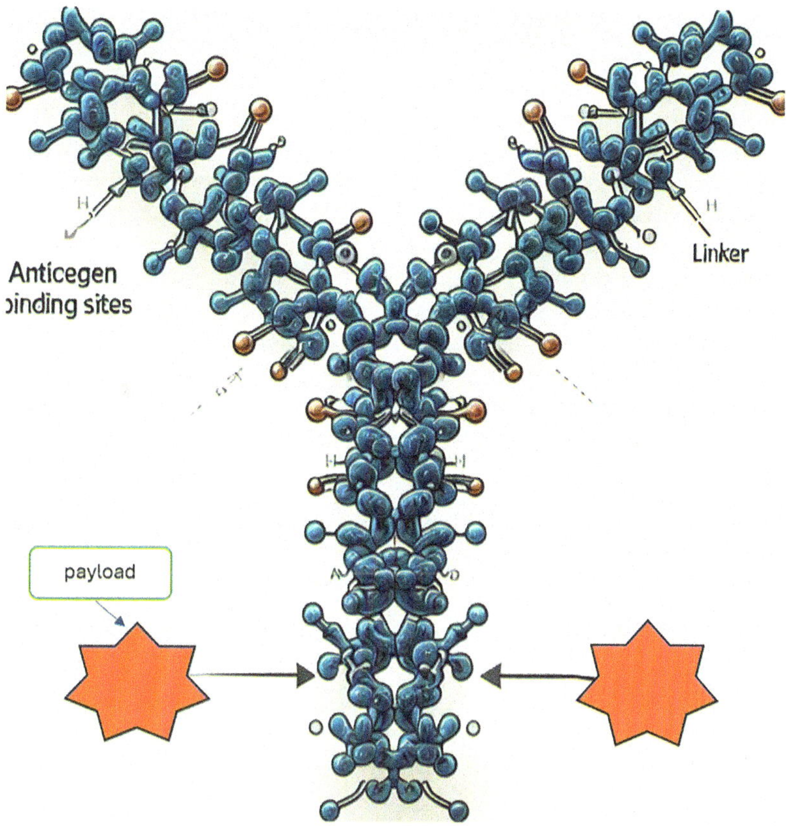

FIG. 4.2 **Parts of an antibody–drug conjugate.** Three parts: antigen-binding site, linker, and cytotoxic payload.

2 **Brentuximab vedotin (Adcetris)**: Used for relapsed Hodgkin lymphoma and systemic anaplastic large-cell lymphoma. It targets CD30-expressing cells.
3 **Inotuzumab ozogamicin (Besponsa)**: Approved for CD22-positive B-cell precursor acute lymphoblastic leukemia that has relapsed or is refractory.

Common ADC targets: HER2, Trop2, Nectin4, EGFR, CD19, CD22, CD33, CD30, BCMA, and CD79b are a few well-known ADC targets. These targets have been successfully used in approved ADCs for various cancers.

Emerging targets

Researchers are exploring newer targets with promising results.

1 **EpCAM**: Epithelial cell adhesion molecule, implicated in several cancers.
2 **CD70**: Overexpressed in certain lymphomas and solid tumors.
3 **CD25**: A potential target in hematological malignancies.

Ongoing research aims to optimize ADC design, minimize toxicity, and enhance antitumor specificity.

Conclusion

Combination therapy has the potential to be more effective in targeting several pathways involved in tumor growth and progression, which is a major advantage in the treatment of cancer. Physicians can limit the risk of drug resistance and increase the overall anticancer benefits by using a variety of medications with distinct mechanisms of action. In addition, combination therapy has shown promise in improving patient outcomes, including higher response rates and prolonged survival. However, challenges in implementing combination therapy include increased toxicity and the potential for drug interactions, which may lead to severe consequences and affect the quality of life for patients. Furthermore, the complexity of managing multiple drugs simultaneously can be burdensome for both patients and healthcare providers. Overall, while combination therapy offers significant advantages in treating cancer, careful consideration of its benefits and challenges is essential for optimizing treatment outcomes.

References

Akhade, A., Van Wambeke, S., Gyawali, B., 2022. CDK 4/6 inhibitors for adjuvant therapy in early breast cancer—Do we have a clear winner? Ecancermed. Sci. 1754-6605. 16. doi:10.3332/ecancer.2022.ed124.

Arruebo, M., Vilaboa, N., Sáez-Gutierrez, B., Lambea, J., Tres, A., Valladares, M., González-Fernández, Á., 2011. Assessment of the Evolution of Cancer Treatment Therapies. Cancers. 2072-6694. 3 (3), 3279–3330. doi:10.3390/cancers3033279.

Cattrini, C., Zanardi, E., Boccardo, F., 2018. Androgen-deprivation therapy is more than palliation in oligometastatic prostate cancer. J. Clin. Oncol. 1527-7755. 36 (22), 2350. Available from: http://ascopubs.org/doi/pdf/10.1200/JCO.2018.78.0031. doi:10.1200/JCO.2018.78.0031.

DeForge, S.M., Baltazar, A.R., Enzinger, A.C., Edwards, R.R., Tulsky, J.A., Schreiber, K.L., Allsop, M., Businelle, M., Azizoddin, D.R., 2023. Perspectives of patients with advanced cancer on mobile health interventions to support coping with cancer pain. JCO Oncol. Pract. 2688-1527. 19 (11_suppl), 277. doi:10.1200/op.2023.19.11_suppl.277.

Delou, J.M.A., Souza, A.S.O., Souza, L.C.M., Borges, H.L., 2019. Highlights in resistance mechanism pathways for combination therapy. Cells 2073-4409. 8 (9). Available from: https://www.mdpi.com/2073-4409/8/9/1013/pdf. doi:10.3390/cells8091013.

Doostmohammadi, A., Jooya, H., Ghorbanian, K., Gohari, S., Dadashpour, M., 2024. Potentials and future perspectives of multi-target drugs in cancer treatment: the next generation anti-cancer agents. Cell Commun. Signal. 1478-811X. 22 (1). doi:10.1186/s12964-024-01607-9.

Ezike, T.C., Okpala, U.S., Onoja, U.L., Nwike, C.P., Ezeako, E.C., Okpara, O.J., Okoroafor, C.C., Eze, S.C., Kalu, O.L., Odoh, E.C., Nwadike, U.G., Ogbodo, J.O., Umeh, B.U., Ossai, E.C., Nwanguma, B.C., 2023. Advances in drug delivery systems, challenges and future directions. Heliyon 2405-8440. 9 (6). Available from: http://www.journals.elsevier.com/heliyon/. doi:10.1016/j.heliyon.2023.e17488.

Garg, P., Malhotra, J., Kulkarni, P., Horne, D., Salgia, R., Singhal, S.S., 2024. Emerging Therapeutic Strategies to Overcome Drug Resistance in Cancer Cells. Cancers 2072-6694. 16 (13). Available from: http://www.mdpi.com/journal/cancers/. doi:10.3390/cancers16132478.

Gerhardtova, I., Cizmarova, I., Jankech, T., Olesova, D., Jampilek, J., Parrak, V., Nemergutova, K., Sopko, L., Piestansky, J., Kovac, A., 2024. Implementation of Modern Therapeutic Drug Monitoring and Lipidomics Approaches in Clinical Practice: a Case Study with Colistin Treatment. Pharmaceuticals 1424-8247. 17 (6). Available from: https://www.mdpi.com/journal/pharmaceuticals. doi:10.3390/ph17060753.

Hayashi, N., Niikura, N., Yamauchi, H., Nakamura, S., Ueno, N.T., 2013. Adding hormonal therapy to chemotherapy and trastuzumab improves prognosis in patients with hormone receptor-positive and human epidermal growth factor receptor 2-positive primary breast cancer. Breast Cancer Res. Treat. 1573-7217. 137 (2), 523–531. doi:10.1007/s10549-012-2336-6.

Huo, G., Liu, W., Kang, S., Chen, P., 2023. Toripalimab plus chemotherapy vs. chemotherapy in patients with advanced non-small-cell lung cancer: a cost-effectiveness analysis. Front. Pharmacol. 1663-9812. 14. doi:10.3389/fphar.2023.1131219.

Liu, K., Yang, Q., Liu, P., Zhu, K., Zou, M., Zhu, Q., Yi, P., Fang, K., Luo, Z., Bertolini, F., 2024. CD70 is a potential prognostic marker and significantly regulates cellular function in diffuse large B-cell lymphoma. PLOS ONE. 1932-6203. 19 (10), e0312445. doi:10.1371/journal.pone.0312445.

Mokhtari, R.B., Homayouni, T.S., Baluch, N., Morgatskaya, E., Kumar, S., Das, B., Yeger, H., 2017. Combination therapy in combating cancer. Oncotarget 1949-2553. 8 (23), 38022–38043. Available from: http://www.impact-journals.com/oncotarget/index.php?journal=oncotarget&page=article&op=download&path%5B%5D=16723&path%5B%5D=53476. doi:10.18632/oncotarget.16723.

Mondelo-Macía, P., García-González, J., León-Mateos, L., Castillo-García, A., López-López, R., Muinelo-Romay, L., Díaz-Peña, R., 2021. Current Status and Future Perspectives of Liquid Biopsy in Small Cell Lung Cancer. Biomedicines. 2227-9059. 9 (1), 48. doi:10.3390/biomedicines9010048.

MU, Y.A.K.U.I., SA, N.A., YU, L.I.A.N.G., LU, S.U.M.E.I., WANG, H.A.I.B.O., XU, W.E.I., 2014. Epithelial cell adhesion molecule is overexpressed in hypopharyngeal carcinoma and suppresses the metastasis and proliferation of the disease when downregulated. Oncol. Lett. 1792-1074. 8 (1), 175–182. doi:10.3892/ol.2014.2140.

Nitiss, J.L., 2009. Targeting DNA topoisomerase II in cancer chemotherapy. Nat. Rev. Cancer. 1474-1768. 9 (5), 338–350. doi:10.1038/nrc2607.

Peng, Y., Tao, Y., Zhang, Y., Wang, J., Yang, J., Wang, Y., 2023. CD25: a potential tumor therapeutic target. Int. J. Cancer 1097-0215. 152 (7), 1290–1303. Available from: http://onlinelibrary.wiley.com/journal/10.1002/(ISSN)1097-0215. doi:10.1002/ijc.34281.

Peters, C., Brown, S., 2015. Antibody-drug conjugates as novel anti-cancer chemotherapeutics. Biosci. Rep. 1573-4935. 35 (4). Available from: http://www.bioscirep.org/content/ppbioscirep/35/4/e00225full.pdf. doi:10.1042/BSR20150089.

Tolcher, A., Hamilton, E., Coleman, R.L., 2023. The evolving landscape of antibody-drug conjugates in gynecologic cancers. Cancer Treat. Rev. 0305-7372. 116, 102546. doi:10.1016/j.ctrv.2023.102546.

Vanneman, M., Dranoff, G., 2012. Combining immunotherapy and targeted therapies in cancer treatment. Nat. Rev. Cancer. 1474-175X. 12 (4), 237–251. doi:10.1038/nrc3237.

Welsch, E., Holzer, B., Schuster, E., Fabikan, H., Weinlinger, C., Hauptmann-Repitz, E., Illini, O., Hochmair, M.J., Fischer, M.B., Weiss, E., Zeillinger, R., Obermayr, E., 2024. Prognostic significance of circulating tumor cells and tumor related transcripts in small cell lung cancer: a step further to clinical implementation. Int. J. Cancer 1097-0215. 154 (12), 2189–2199. Available from: http://onlinelibrary.wiley.com/journal/10.1002/(ISSN)1097-0215. doi:10.1002/ijc.34886.

Zheng, Z., Chen, H., Cai, H., 2024. Cost-effectiveness analysis of serplulimab combination therapy versus chemotherapy alone for patients with extensive-stage small cell lung cancer. Front. Oncol. 2234-943X. 13. doi:10.3389/fonc.2023.1259574.

CHAPTER

5

Drug resistance: Mechanisms and strategies to overcome

Bipasha Kalita[*,a], *Sankar Jyoti Bora*[*,a], *Phulen Sarma*[a], *and Sadhna Aggarwal*[b]

[a]Department of Pharmacology, AIIMS, Guwahati, India

[b]Department of Radiation Oncology, University of Texas MD Anderson Cancer Center, Houston, TX, United States

Introduction

A condition known as drug resistance occurs when a disease becomes resistant to medication. The problem of drug resistance is not only limited to infectious diseases but also extends its impact to the field of cancer (Gottesman, 2002). The development of drug resistance has rendered the treatment of cancer ineffective, posing a new treatment challenge. Cancer remains a significant disease burden across the entire world. One of the biggest health issues and socioeconomic burdens of the 21st century is cancer. Noncommunicable diseases such as cardiovascular diseases and cancer are leading causes of death worldwide. Cancer is the second most common cause of mortality worldwide, after heart disease (Ma and Yu, 2006; Boutayeb, 2010). According to the International Agency for Research on Cancer's (IARC) GLOBOCAN 2022 study, 20 million new cases of cancer were reported in 2022, and by 2050, there will likely be 35 million cases globally. Numerous factors, including inherited genetic traits, exposure to environmental toxins, poor lifestyle choices, and an aging population, contribute to the rising number of cancer cases worldwide. Although modern treatments such as immunotherapy and targeted therapy have improved cancer care, drug resistance remains a significant barrier, frequently resulting in treatment failure and disease progression (Bray et al., 2024).

The majority of cancer treatments vary depending on the patient, the type of disease, its stage, molecular features, and the patient's general health. Eliminating cancer cells

*Equal contribution.

Pharmacological Approaches in Cancer Treatment
https://doi.org/10.1016/B978-0-443-38307-6.00004-1

while causing the least amount of damage to healthy tissue is the main aim, achieved through modalities such as surgery, chemotherapy, targeted therapy, radiation therapy, endocrine therapy, and immunotherapy (Liu et al., 2024; Gale, 2024). Surgery is mostly preferred in the early stages, where tumors can be surgically removed. Common examples include mastectomy in breast cancer and colectomy in colorectal cancer (CRC) (Osborne and Borgen, 1990; Renkonen-Sinisalo et al., 2017). Chemotherapy is a widely used and promising treatment, particularly in advanced-stage cancer, mainly utilizing cytotoxic drugs to kill or inhibit cancer cell growth, often in combination with surgery, radiation therapy, or targeted therapy. Radiation therapy uses high-energy radiation to destroy cancer cells. It is commonly used for localized tumors postoperatively to prevent recurrence, and as palliative radiotherapy (Brunner, 2016). Targeted therapies, such as trastuzumab for Human Epidermal Growth Factor Receptor 2 (HER-2)-positive breast cancer, specifically attack cancer cells based on molecular features (Mercogliano et al., 2023). Immunotherapies act by stimulating or modifying the immune response, such as checkpoint inhibitors pembrolizumab and nivolumab (Wong et al., 2021). By preventing the activity or synthesis of hormones, endocrine (hormone) therapy works for tumors that are sensitive to hormones, such as breast and prostate cancer. Tamoxifen is an example of this treatment for estrogen receptor (ER)-positive breast cancer (Criscitiello et al., 2010). Despite significant breakthroughs in chemotherapy, curative medicines for late-stage malignancies are largely unavailable due to chemoresistance. Combination therapies have improved chemotherapeutic results by reducing chemoresistance and boosting chemosensitivity.

Drug resistance in cancer can be divided into two broad categories. Intrinsic drug resistance is due to factors present in the cancer cells that make the cancer resistant or ineffective to chemotherapy, whereas extrinsic or acquired drug resistance can develop during the treatment of tumors where the tumors were initially sensitive to chemotherapy but later become resistant. Drug efflux mechanisms, drug inactivation and detoxification, drug target alteration, DNA damage repair mechanisms, apoptosis evasion, epigenetic modification, tumor microenvironment (TME) influence, autophagy-mediated resistance, and immune evasion are some of the key mechanisms of chemoresistance, as depicted in Fig. 5.1 (Mansoori et al., 2017; Khan et al., 2024). Drug resistance has been observed with doxorubicin, cyclophosphamide, 5-fluorouracil, paclitaxel, cisplatin, and other medications used to treat the majority of deadly malignancies, including colorectal, lung, and breast cancers, among others (Christowitz et al., 2019; Ludeman and Gamcsik, 2002; Sethy and Kundu, 2021; ER et al., 2022; Shen et al., 2012). To understand the molecular aspects of this resistance, numerous studies have been conducted. Given that cancer chemoresistance causes cancer recurrence, metastatic progression, and overall treatment failure, these data unmistakably demonstrate a substantial disease burden. Drug resistance increases the burden of disease, which raises mortality and healthcare costs due to longer hospital stays (Fongang et al., 2024). To effectively manage medication resistance in cancer, customized treatment, combination therapies, and innovative drug discovery are essential. In addition, understanding the mechanisms of chemoresistance is crucial for developing new strategies to overcome resistance and improve the effectiveness of cancer treatments. Furthermore, in this chapter, different promising strategies to overcome these mechanisms of resistance have also been explained.

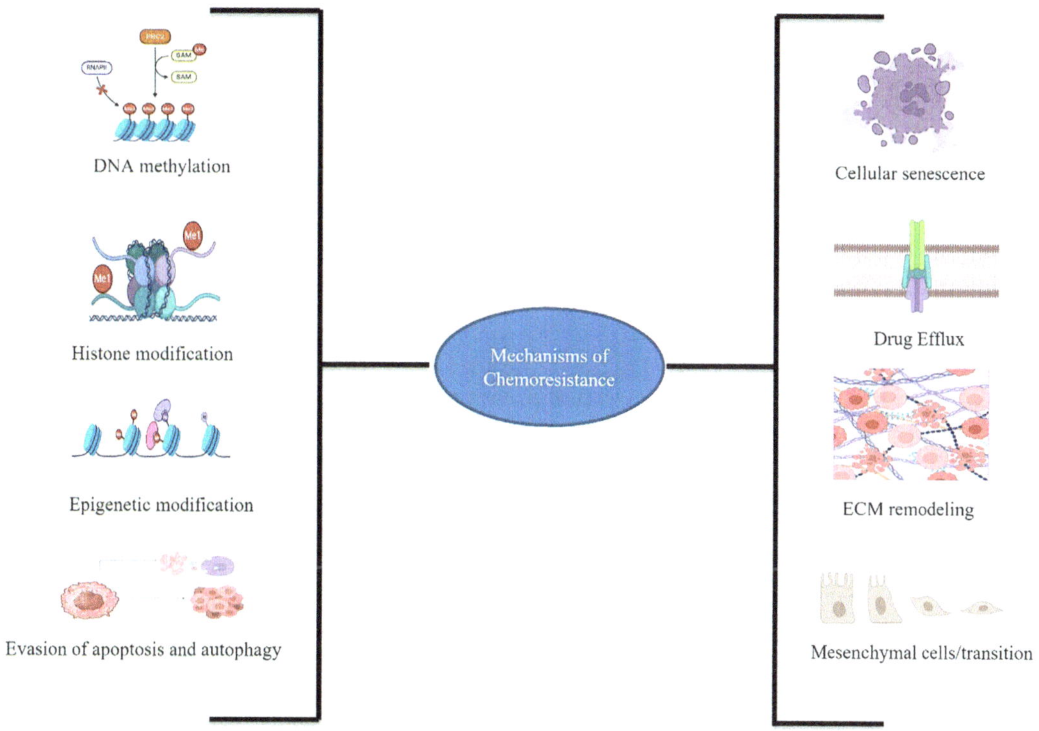

FIG. 5.1 Mechanisms of chemoresistance.

Abnormal drug transport pathways

One key mechanism behind chemoresistance involves abnormal drug transport pathways. These pathways alter how drugs enter, move within, and exit cancer cells, reducing the drugs' ability to kill the cancer cells. By changing drug efflux, uptake, sequestration, and detoxification, P-glycoprotein (P-gp), SLC31A1, ATP7A, ATP7B, metallothionein (MT), and glutathione (GSH) contribute to aberrant drug transport, which results in decreased intracellular drug accumulation and treatment resistance. The distribution, absorption, and excretion of a variety of chemical compounds are regulated by the proteins present in the cell membrane belonging to the ATP-binding cassette, which mainly includes P-gp/ATP-binding cassette subfamily B member 1 (ABCB1) or breast cancer resistance protein. P-gp is a protein found in large amounts on the surface of endothelial cells, which are the cells lining blood vessels. It acts like a pump, pushing substances—including chemotherapeutic drugs—out of the cells. When P-gp is highly active on the cells of the BBB, it further reduces the ability of chemotherapy drugs to cross into the brain (Bukowski et al., 2020). In resistant ovarian cancer cells that have high levels of P-gp (called "overexpression") encoded by ABCB1, this pump works to push chemotherapy drugs like paclitaxel, doxorubicin, and sorafenib out of the cell before they can do their job—killing the cancer—leading to chemoresistance (Wang et al., 2024b). More recently, researchers have found that some drugs originally developed for other cancer treatments (like selonsertib, tepotinib, and poziotinib) can also block both ABCB1 and ABCG2 pumps, weakening MDR mediated by ABCB1 and ABCG2 in cancer cells overexpressing these transporters (Lei et al., 2023). CRISPR/CAS9, which acts like an RNA-based adaptive immune system, has been explored to overcome chemoresistance. Using this technology, ABCB1 gene can be targeted and knocked out, thus increasing sensitivity to doxorubicin (Vaghari-Tabari et al., 2022). SLC31A1 is like a "gateway" protein on the surface of cells. It helps to let certain drugs, like cisplatin, carboplatin, and oxaliplatin (which are used in ovarian cancer treatment), enter the cell. Once inside, these drugs accumulate and release platinum, which helps kill the cancer cells. If this "gateway" stops working (this is called downregulation), fewer drugs can enter the cell, resulting in ineffective accumulation of these drugs inside to fight against the cancer (Wang et al., 2024b). SLC31A1 is also a cuproptosis-related gene. Intracellular copper accumulation contributes to the aggregation of mitochondrial lipoylated proteins, leading to a novel form of programmed cell death termed cuproptosis. Dysregulation of cuproptosis-related genes may lead to tumor progression (Wu et al., 2023). ATP7A and ATP7B are P-type ATPases referred to as copper-transporting ATPases because they move copper ions out of cells. They keep the efflux and homeostasis of copper stable. Their overexpression may help ovarian cancer cells release platinum medications like cisplatin (Li et al., 2018). MT and GSH are intracellular thiol-containing proteins that are rich in cysteine. Cisplatin, which is present in the cytoplasm, binds to those proteins, leading to its sequestration, which is a major hurdle to overcome. By attaching to metal-based medications such as cisplatin and blocking their ability to interact with DNA, MTs contribute significantly to drug resistance by decreasing the treatments' capacity to eradicate cancer cells. Their role in controlling oxidative stress further promotes cancer cell survival, and their increased expression in some malignancies has been connected to resistance to cisplatin. Similarly, a crucial antioxidant called GSH aids in drug detoxification and neutralization of dangerous reactive oxygen species (ROS). GSH encourages the clearance of medicines from cells by building complexes with them via glutathione S-transferase, which reduces the efficacy of the medications. Drug-resistant cancer cells with higher GSH levels are more resistant to chemotherapy because they offer more defense against oxidative stress (Lee et al., 2015).

Cancer stem cells

A tiny minority of cancer cells, called cancer stem cells (CSCs), can self-renew and differentiate inside a tumor, which may contribute to medication resistance and tumor development, recurrence, and metastasis. CSCs maintain their self-renewal capacity through cellular plasticity, senescence, and quiescence. Stemness is maintained by plasticity and genetic reprogramming, while stemness activation is induced by senescence, and therapeutic stress is overcome by the state of quiescence, ultimately leading to therapy resistance (Paul et al., 2022). The overexpression of CSC markers, such as CD24 or MET, is associated with resistance to platinum-based chemotherapy in ovarian cancer patients (Kwon et al., 2024). Overexpression of ZEB2, a transcription factor, has been linked to an increase in the proportion of colorectal CSCs in the G0/G1 phase, which is a quiescent state (Chu et al., 2024). Mechanoregulator proteins, such as integrin, modulate the characteristics of tumor stem cells through different molecular signaling pathways, and these proteins are mainly activated by extracellular matrix (ECM) stemness

(Safaei et al., 2023). This higher matrix stiffness has been reported to promote the CSC phenotype and reduce sorafenib-induced apoptosis in HCC. Epithelial cell adhesion molecule ($EpCAM^+$) is also a commonly used biomarker for liver CSC isolation (Niu et al., 2024). Mesenchymal stem/stromal cells (MSCs) are stromal cells in the TME, and cancer cells interact with them, resulting in the transformation of naïve MSCs to tumor-associated MSCs. This transformation has an important impact on tumor growth, progression, and the development of chemoresistance in cancer cells (Ramuta and Kreft, 2022). It has been found that tumor cells evade immune attacks with the help of PGE2 secreted by the MSCs in response to tumor cell-derived interleukin-1. CSCs' growth and maintenance of their stem-like characteristics are facilitated by IL-6 and hepatocyte growth factor secreted by MSCs (Guo et al., 2023). Mesenchymal stem cells promote epithelial-to-mesenchymal transition, CSC propagation, evasion of the immune system, and enhancement of therapeutic resistance in breast cancer patients (Tu and Karnoub, 2022).

SLC7A11, stress, ferroptosis, and cancer therapy

SLC7A11 acts as an important oncogenic protein. It reduces oxidative stress with the help of GSH and regulates many pathophysiological processes, such as ferroptosis and drug resistance in cancer. SLC7A11 becomes highly active in certain cancer cells, helping them combat oxidative stress and survive (Lin et al., 2020). Ferroptosis is a regulated form of cell death caused by excessive iron accumulation and cellular damage from lipid peroxidation, essentially "rusting" to the point where they can no longer function. The upregulation or activation of SLC7A11, which reduces sensitivity to ferroptosis, can enhance the fitness of cancer cells, promoting proliferation, metastasis, and chemoresistance (Koppula et al., 2021). In patients with KRAS-mutant lung adenocarcinoma (a type of lung cancer), the SLC7A11 protein is produced in very high amounts. This overproduction is associated with tumor growth and progression. By targeting SLC7A11 with specific drugs, such as sulfasalazine, erastin, and imidazole ketone erastin, scientists aim to block its transportation activity. This inhibition reduces the levels of GSH in cancer cells, making them more vulnerable to damage and death, particularly through processes like ferroptosis (Su et al., 2025). Some drugs target SLC7A11 differently, not by blocking its activity, but by reducing its production. This can be achieved at the mRNA level or the protein level. For instance, microRNA (miRNA) negatively regulates SLC7A11 by decreasing its expression in a type of bladder cancer that is resistant to the chemotherapy drug cisplatin (Drayton et al., 2014). Metformin increases intracellular iron and lipid ROS levels, destabilizing the SLC7A11 protein. When metformin is combined with another drug, sulfasalazine, they work synergistically to amplify the effect. This combination triggers ferroptosis. By inducing ferroptosis in cancer cells, this drug duo helps shrink tumors in a breast cancer model (Yang et al., 2021).

Apoptosis evasion

Autophagy involves the degradation of cytoplasmic proteins and organelles via the lysosomal pathway, with the help of autophagy-related genes (ATGs). It proceeds in a stepwise manner: initiation, nucleation of the autophagosome, expansion and elongation of the autophagosome membrane, closure and fusion with the lysosome, and the degradation of intravesicular products (Xie et al., 2020). Autophagy, in its initial stages, can lead to tumor cell growth inhibition by enhancing the activity of tumor suppressor genes, but in the advanced stages, it can also lead to the progression of tumor development by stimulating tumor growth; thus it can be considered a double-edged sword. ATGs are involved in the regulation of the apoptotic pathway. The ATG12–ATG3 complex has an autophagy-independent function that can regulate apoptosis. ATG3 mutations prevent conjugation to ATG12 and, through the mitochondrial pathway, inhibit apoptosis in an autophagy-independent manner, thereby highlighting that these two pathways are interconnected (Vitto et al., 2022). When treatments like radiation or chemotherapy damage DNA, cells try to repair it. This response can unintentionally activate autophagy, giving cancer cells more tools to resist treatment.

- The ataxia telangiectasia mutated (ATM) protein is a key player in the DNA damage response (DDR). It senses DNA damage and activates repair mechanisms. One of these mechanisms triggers autophagy, which helps cancer cells survive.
- The homologous recombination (HR) repair pathway, another repair system activated by ATM, can further support autophagy (Khan et al., 2024).

Cellular autophagy can be inhibited by the lysosomal inhibitor bafilomycin A1, which enhances epirubicin-triggered apoptosis due to the inhibition of autolysosome formation, ultimately leading to the suppression of autophagic effects such as engulfing and clearing damaged mitochondria. Mitochondrial cytochrome C release is increased by this inhibition, which augments epirubicin-induced caspase-dependent apoptosis and cytotoxicity in breast cancer cells (Guo et al., 2016). In healthy cells, the Bcl-2 family proteins regulate apoptosis, the programmed self-destruction process that removes damaged or unnecessary cells. The Bcl-2 family is divided into two main groups: proapoptotic proteins, which promote apoptosis and help eliminate unhealthy cells (examples include BAX and BAK), and antiapoptotic proteins, which prevent apoptosis and support cell survival (examples include Bcl-2 and Bcl-xL). In cancer, this delicate balance is disrupted, favoring cell survival. Several factors, such as genetic mutations, gene amplifications, and aberrant protein expression, contribute to this imbalance, resulting in increased amounts of antiapoptotic proteins and reduced production of pro-apoptotic proteins (Tufail et al., 2024). Apoptosis is regulated by inhibitor of apoptosis proteins (IAPs), which are a family of proteins that inhibit caspases. Caspases are cysteine-dependent aspartate-directed proteases that serve as the effector arm of the apoptosis process (Kumar et al., 2020a). Structurally distinct IAPs—X-linked (XIAP), cellular (cIAP1, cIAP2), neuronal (NIAP), testis-specific (Ts-IAP), Bir-ubiquitin conjugating enzyme (BRUCE), Survivin, and Livin—modulate the execution of extrinsic and intrinsic apoptotic signals. IAPs are approximately 70 amino acids long and contain zinc finger Baculovirus IAP Repeat (BIR) domains, which contribute to the inhibitory properties of IAPs by preventing the conversion of zymogenic procaspases to active caspases (Rathore et al., 2017). For example, elevated levels of XIAP in ovarian cancer patients have been linked to resistance against platinum-based chemotherapy, such as carboplatin. XIAP inhibition leads to repression of cell viability in carboplatin-treated cells and increased carboplatin-induced caspase activation. Thus XIAP mediates carboplatin sensitivity in ovarian cancer and helps cancer cells survive and resist the effects of therapy (Zhang et al., 2018). The p53 protein is encoded by the TP53 gene in humans and is known as the "guardian of the genome" because it maintains genomic stability through the repair of damaged DNA and the integration of DNA damage checkpoints with cell birth/death signaling pathways. p53 can regulate diverse cellular processes, including DNA repair, cell cycle, apoptosis, redox homeostasis, and metabolism under physiological or various cellular stress conditions (Park et al., 2016). Mutated p53, with reduced or abolished function, is often overexpressed and connected to resistance to several chemotherapeutic drugs, including cisplatin, temozolomide, doxorubicin, gemcitabine, tamoxifen, and EGFR inhibitors such as cetuximab. Changes in the conformation of the p53 protein are often accompanied by mutations in the TP53 gene (Hientz et al., 2017).

The death receptor pathway is another major mechanism that activates apoptosis through "death receptors" such as FAS, TRAIL-R1, and TRAIL-R2, which are present on the surface of cells.

Cancer cells can evade this pathway by:

- Modifying the expression of death receptors: They may reduce the number of these receptors on their surface, making it harder for the apoptosis signal to reach them.
- Altering the localization of receptors: Cancer cells may change the location of these receptors, moving them away from areas where they would normally be activated.
- In addition, it has been discovered that cancer cells can manipulate the "trafficking" (movement) of death receptors inside the cell, which is a novel and advanced way to block the signals that would trigger apoptosis (Roszkowska, 2024).

Epigenetic modifications and the role of micro-RNAs

Epigenetic processes, which are reversible and dynamic, primarily regulate changes in chromatin structure. In the higher-order folding of chromatin fibers, nuclear remodeling responds to changes in epigenetic modifications, which include DNA methylation, histone modification, and RNA-mediated processes. Histone tail modifications act as expression or repression markers through the acetylation and methylation of various amino acids. Epigenetic codes, in the form of DNA methylation or posttranslational histone modifications, are stored and transmitted by chromatin. During cell replication, these modifications are inherited; thus epigenetic dysregulation is a feature of nearly all human cancers (Yu et al., 2024). Noncoding RNAs include miRNAs, short interfering RNAs (siRNAs), and long noncoding RNAs (lncRNAs). Both miRNAs and lncRNAs interact with numerous epigenetic modifiers and transcription factors, influencing the expression of genes. The development and progression of tumors are related to the aberrant expression of these ncRNAs. Dysregulation of ncRNAs is caused by epigenetic alterations. Most abnormally expressed ncRNAs alter cellular proliferation and apoptotic pathways, resulting in cancer progression and treatment resistance. Proliferation and progression of cancer can be prevented by either the upregulation or downregulation of ncRNAs (Kumar et al., 2020b). Regulation of ferroptosis in cancer is modulated by miRNAs. In hepatocellular carcinoma (HCC), miR-214-3p promotes ferroptosis by downregulating the expression of activating transcription factor 4 (ATF4). Stress signals induce ATF4, and thus, ferroptosis is prevented through the induction of SLC7A11 (Balihodzic et al., 2022). In cisplatin-resistant A549 (A549/DDP) cells, downregulated expression of miR-324-3p occurs. GPX4, an antioxidant enzyme involved in the regulation of ferroptosis, is targeted by miR-324-3p, and overexpression of GPX4 reverses miR-324-3p-mediated increased sensitivity of A549/DDP cells to cisplatin. RSL3, which is a GPX4 inhibitor, mimics the effects of overexpressed miR-324-3p in increasing the sensitivity of cisplatin-resistant NSCLC cells to the drug (Wang et al., 2024a).

DNA damage response

Genomic integrity is secured by a complex and intricate network of DDR mechanisms, including multiple DNA repair pathways, damage tolerance processes, and cell-cycle checkpoints. DDR is a chromatin-associated process that is generally tightly controlled in time and space, similar to transcription and replication. A specialized spatiotemporal orchestration of this defense apparatus is required, as DNA damage can occur at any time and at any genomic location (Giglia-Mari et al., 2011). DNA alteration contributes to therapy resistance through genomic instability and the associated production of new antigens, as well as the upregulation of PD-L1 (programmed death ligand 1), which interacts with signaling molecules such as cyclic GMP–AMP synthase–stimulator of interferon genes (cGAS–STING) (Jiang et al., 2021). PD-L1 interacts with the PD-1 receptor on immune cells (such as T cells) and essentially turns off the immune response. By increasing PD-L1, cancer cells create a shield that allows them to evade immune attack. This is especially problematic in immunotherapy, where the goal is to boost the immune system's ability to recognize and kill cancer cells (Jiang et al., 2019). In response to DNA damage, there is rapid activation of checkpoint kinases. Checkpoint kinases are signaling cascades that result in cell cycle arrest, allowing time for DNA repair or the induction of apoptotic cell death. Conserved checkpoint protein kinases, including ATM, ATM- and Rad3-related (ATR) proteins, and the downstream checkpoint kinases CHK2 and CHK1 carry out the transduction of the DDR signal induced by 5-FU in colorectal cancer patients. Phosphorylation of CHK1 at S317 by the ATR kinase is required for its activation in response to DNA damage. CHK1 is upregulated in CRC, and its overexpression and/or hyperactivation leads to chemo- and radiotherapy resistance (Yu et al., 2020). Cells are equipped with specialized enzymes, including apurinic/apyrimidinic (AP) endonuclease 1 (APE1), polynucleotide kinase-phosphatase (PNKP), tyrosine DNA phosphodiesterase 1 (TDP1), aprataxin (APTX), and DNA ligases, to prevent the harmful effects of DNA single-strand break repair (SSBR) damage, which is one of the most common DNA lesions. In addition, several other kinases related to cell vulnerability, such as breast cancer susceptibility gene 1 (BRCA1), radiation-sensitive 51 (RAD51), RAD52, xeroderma pigmentosum group C, and flap endonuclease 1 (FEN1), are involved in the DDR (Gao et al., 2024). ATM is a critical protein in the DDR system, specifically responding to double-strand breaks (DSBs) in DNA. These breaks are one of the most severe types of DNA damage and can lead to genomic instability or cell death if not properly repaired. Abnormal ATM activation is often one of the key factors through which tumor cells develop tolerance to treatment. To exploit cancer cells' inherent vulnerability to DNA damage, various small molecule inhibitors have been developed that target ATM, which have shown great versatility in preclinical studies and increasing use in the clinic (Ampolini et al., 2025).PARP is a family of nuclear protein enzymes involved in DDR. Poly (ADP-ribose) polymerase 1 (PARP1) acts as a sensor for damaged nucleotides. It is a protein located in the nucleus in high abundance. Following SSB or DSB DNA damage, PARP1 enzymatic activity is increased, leading to the synthesis of poly-ADP-ribose (PAR) chains after damage. The modification of proteins such as DNA polymerases, topoisomerases, and DNA ligase-2, as well as histones, high-mobility-group proteins, and transcription factors, allows PARP1 to control not only cellular repair, DNA replication, and transcription but also protein degradation, organization of the cytoskeleton, and other cellular functions (Jurkovicova et al., 2022). Inhibition of PARP impairs the repair of SSBs, leading to synthetic lethality in HR-deficient (such as deleterious BRCA1/2) cells, which are unable to accurately repair DSBs. Several PARP inhibitors have been approved by the FDA for use against different cancers, with or without HR deficiency, such as olaparib, niraparib, rucaparib, and talazoparib. Studies on PARP1/2 inhibitors as monotherapy have proposed them as a novel therapeutic strategy for NSCLC cells (Mao et al., 2024).

Tumor microenvironment

The cellular environment in which a tumor exists is called TME, consisting of tumor cells, surrounding blood vessels, the ECM, and other nonmalignant cells. It also contains signaling molecules that facilitate cell-to-cell communication. The TME includes not only stromal cells, fibroblasts, and immune cells [such as T lymphocytes, B lymphocytes, natural killer T cells, and tumor-associated macrophages (TAMs)] but also pericytes and sometimes adipocytes. An important role in tumor growth and the suppression of antitumoral immunity is played by the interactions between the tumor cells and molecules released in TME. Because of its dynamic nature, the TME may change over the course of cancer progression, indicating its role in metastasis (Wilczyński et al., 2024). TAMs are considered to be one of the most closely associated tumor-infiltrating immune cells in the tumor cell and TME interaction, as depicted in Fig. 5.2. Macrophages have a high degree of plasticity and can be activated into M1 (antitumor) or M2 (tumor-promoting) phenotypes depending on different microenvironmental stimuli such as hypoxia, ROS, interferon-gamma (IFN-γ), and tumor necrosis factor-alpha (TNF-α). Therefore modulating TAMs and reshaping the immune microenvironment to improve therapy outcomes in CRC is a feasible strategy. M2 (tumor-promoting) phenotypes suppress T-cell activity and secrete immunosuppressive cytokines (Zou et al., 2024). Moreover, the loss of miR-146b promotes the development of TAMs toward the M2 phenotype, thus inhibiting T-cell infiltration and facilitating the transformation of an immune-suppressive TME. Upregulating miR-146b enhances the antitumor activity of immunotherapy. By influencing the stability and translation of messenger RNAs, the miRNA miR-146b plays a critical role in posttranscriptional gene regulation. It is also implicated in inflammatory responses and may contribute to the development of cancer (Zou et al., 2024). Tumor cells in the TME can adapt to the complex microenvironment by reprogramming cholesterol metabolism. Cholesterol, an essential component of the cell membrane, plays a critical role in the survival of immune and tumor cells. Tumor progression and antitumor immune responses can be impacted by the reprogramming of cholesterol metabolism in both tumor cells and immune cells. Strategies aimed at modulating cholesterol metabolism have been demonstrated to be effective in hindering tumor growth and boosting antitumor immune functions (Guo et al., 2025). Statins (HMGCR inhibitors) have been extensively studied as repurposed drugs for their antitumor activity. Statins can exert antitumor properties in vivo

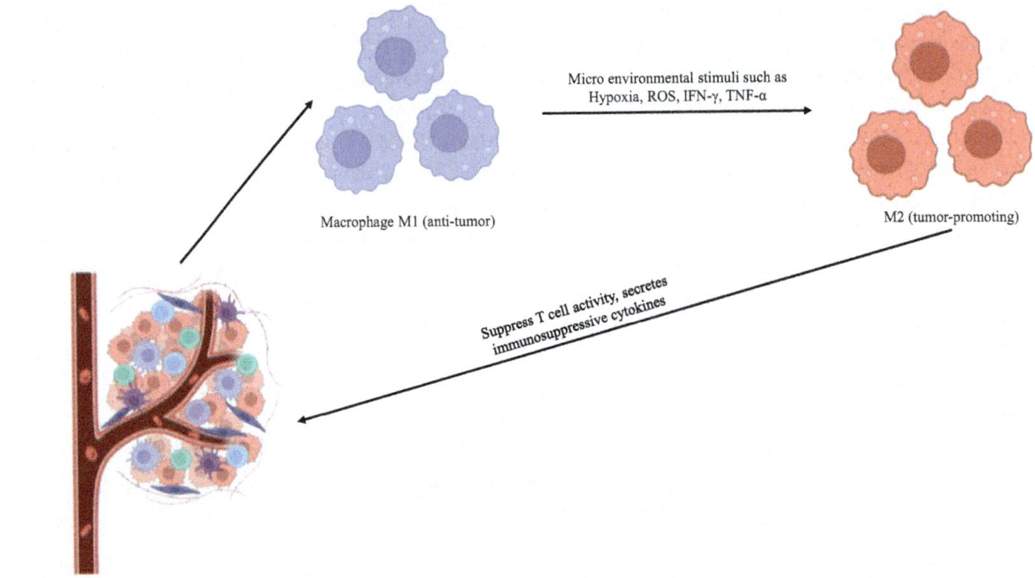

FIG. 5.2 Immune-suppressive tumor microenvironment: The TME's dynamic nature may cause it to alter as the cancer spreads, revealing its involvement in metastasis (Safaei et al., 2023). In the interaction between tumor cells and tumor-associated macrophages (TAMs), TAMs are thought to be among the immune cells that infiltrate tumors most closely. Depending on several micro environmental stimuli, including hypoxia, ROS, interferon-gamma (IFN-γ), and tumor necrosis factor alpha (TNF-α), macrophages can be triggered into either M1 (antitumor) or M2 (tumor-promoting) phenotypes. M2 (tumor-promoting) phenotypes release immunosuppressive cytokines and inhibit T cell function.

and in vitro through multiple mechanisms and have emerged as promising antitumor agents in various clinical trials. A randomized controlled trial has shown that fluvastatin suppresses proliferation and increases apoptosis in high-grade breast cancer. As statins have the ability to downregulate miRNAs known to decrease cholesterol, they have demonstrated possible anticancer properties (Jiang et al., 2024).

Novel drug development for combating chemoresistance

By improving the efficacy and safety of treatments, novel drug delivery systems (NDDSs) are revolutionizing the treatment of cancer. These cutting-edge methods, such as liposomes and nanoparticles, help medications more accurately target cancer cells while minimizing adverse effects (Chen et al., 2024). Overcoming chemoresistance, a significant obstacle that arises when cancer cells cease reacting to conventional therapies, is one of their greatest advantages. NDDSs can increase the effectiveness of treatment by enhancing drug absorption, avoiding resistance mechanisms, improving tumor penetration, decreasing drug degradation, and bypassing efflux pumps. These novel delivery techniques provide fresh promise for treating tumors that are difficult to manage, especially in light of advancements in nanotechnology and personalized medicine (Liu et al., 2021). Some of the commonly formulated nanoparticles for cancer treatment are depicted in Fig. 5.3.

The liposome's aqueous and lipid phases allow for the entrapment of both hydrophilic and lipophilic substances, offering the opportunity to deliver a wide range of active pharmaceutical ingredients with remarkably different chemical compositions and physical characteristics. Hydrophilic and hydrophobic medications are encapsulated in spherical vesicles called liposomes, which are composed of lipid bilayers. Through the enhanced permeability and retention effect, liposomes improve drug stability, extend the duration of drug circulation, and enable the selective accumulation of drugs in tumor tissues (Liu et al., 2022; Olusanya et al., 2018). Multidrug resistance (MDR) mechanisms significantly reduce the effectiveness of paclitaxel, a commonly used chemotherapeutic agent for treating various types of cancer. This is particularly true because P-gp is overexpressed, actively pumping paclitaxel out of cancer cells (Pires et al., 2009). Ginsenoside Rg3, a bioactive substance derived from *Panax ginseng*, is added to the formulation to address these issues due to its antitumor, antiangiogenic, and MDR-reversing properties (Kim

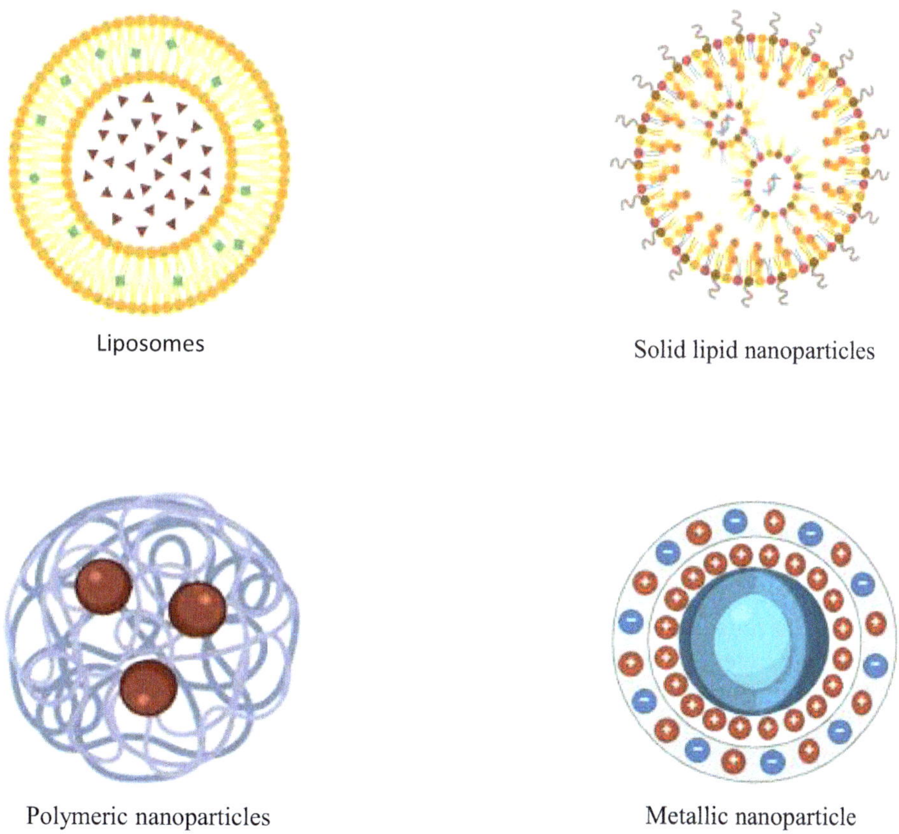

FIG. 5.3 Types of nanoparticles for novel drug development.

et al., 2003). Rg3 has been shown to enhance paclitaxel retention within cancer cells by modulating the TME, suppressing angiogenesis, and downregulating P-gp expression. Furthermore, liposomes reduce systemic toxicity and adverse effects such as neuropathy and hypersensitivity reactions by delivering sustained medication release. In addition, by altering the tumor immunological milieu, Rg3-incorporated liposomes are believed to promote intracellular drug accumulation, prevent drug efflux by P-gp, and enhance chemotherapy response (Zhu et al., 2023). In multidrug-resistant (MDR) malignancies, PEGylated liposomal doxorubicin, a clinically approved formulation, reduces P-gp-mediated efflux and improves therapeutic outcomes. Compared to free doxorubicin, a study on doxorubicin-resistant breast cancer cells (MCF-7/ADR) demonstrated that PEGylated liposomes significantly enhanced intracellular drug accumulation and cytotoxicity (O'Shaughnessy, 2003; Wang et al., 2011).

Biodegradable and biocompatible, solid lipid nanoparticles (SLNs) improve drug stability, release control, and bioavailability. By effectively targeting tumor cells and avoiding drug efflux processes, they reduce systemic toxicity (Mukherjee et al., 2009; Mo et al., 2023). Through a combination of in vitro and in vivo investigations, a study confirmed the chemoresistance mechanism and demonstrated that resveratrol-loaded TPGS-solid lipid nanoparticles (TPGS-SLN) may overcome MDR in breast cancer. The cytotoxic effects of free resveratrol and the nanoparticle formulation were compared in vitro using drug-resistant breast cancer cell lines. By blocking the efflux activity of P-gp, a crucial transporter implicated in medication resistance, TPGS-SLN markedly improved drug retention. A combination of solid lipid carriers, D-α-tocopheryl polyethylene glycol 1000 succinate (TPGS), and resveratrol as the therapeutic ingredient was used to carefully construct the resveratrol-loaded TPGS-SLN. Resveratrol was shielded from degradation and allowed for controlled release by the stable matrix that the solid lipid core provided. The drug's bioavailability was increased by this lipid-based approach, ensuring longer circulation and better therapeutic outcomes. The vitamin E-derivative TPGS is well known for its capacity to block P-gp, a crucial efflux transporter that drives medications out of cancer cells and causes chemotherapy resistance. The formulation successfully decreased drug efflux by incorporating TPGS into the nanoparticle system, which made it possible for resveratrol to accumulate inside cancer cells and more effectively carry out its therapeutic actions (Wang et al., 2021c). Similar improvements were made to drug accumulation and therapeutic performance in resistant cancer models by other formulations, such as paclitaxel–SLNs (Baek et al., 2015; Wang et al., 2021b). In addition, doxorubicin–SLNs have been shown to improve tumor penetration and decrease cardiotoxicity, which suggests that they could be a viable substitute for traditional chemotherapy (Kang et al., 2010).

Because of its mucoadhesive qualities, chitosan, a biocompatible polysaccharide, has been extensively investigated for drug administration (Herdiana et al., 2022). Doxorubicin-loaded chitosan nanoparticles increase chemosensitivity by altering the TME and improving drug retention in cancer cells (Yoncheva et al., 2020). According to recent research, chitosan nanoparticles derived from natural products may improve therapeutic efficacy even further. The study found that paclitaxel-resistant lung cancer cells were successfully sensitized by quercetin-loaded chitosan nanoparticles, increasing their receptiveness to paclitaxel therapy. Due to the enhanced intracellular delivery of quercetin by the chitosan-based nanoparticles, resistant cancer cells were better able to retain the drug. Strong electrostatic interactions with the negatively charged cancer cell membranes were made possible by the cationic nature of the nanoparticles, which facilitated the effective internalization of the drug-loaded nanoparticles. The nanoparticles increased paclitaxel accumulation in the cells by downregulating important resistance-related proteins, including P-gp, which is responsible for drug efflux (Wang et al., 2021d). Curcumin-chitosan nanoparticles demonstrated enhanced bioavailability and anticancer effects in breast cancer xenografts (Deepika et al., 2022).

Drug delivery and biomedical applications frequently use poly(lactic-co-glycolic acid) (PLGA) nanoparticles, which are biodegradable and biocompatible polymeric carriers. PLGA is safe for use in medicine because it spontaneously breaks down into nontoxic byproducts from its monomers, lactic and glycolic acids. These nanoparticles provide continuous and regulated medication release, preventing drug deterioration and increasing bioavailability (Makadia and Siegel, 2011). Furthermore, it has been demonstrated that doxorubicin–PLGA nanoparticles circumvent MDR pathways by promoting intracellular drug delivery and inhibiting drug efflux (Yoo et al., 2000).

Designed for targeted medication delivery and cancer treatment, metallic nanoparticles (MNPs) are nanoscale structures made of metals such as gold, silver, iron, and platinum. By producing ROS, blocking drug efflux pumps, and altering the TME, they aid in the fight against chemoresistance. MNPs increase the effectiveness of chemotherapy against cancer cells that are resistant by promoting drug retention, triggering ferroptosis, and interfering with autophagy (Villalobos Gutiérrez et al., 2023; Rahman et al., 2021). Zinc oxide nanoparticles (ZnO-NPs) efficiently prevent the growth, migration, and invasion of gastric cancer (GC) cells while triggering apoptosis. Notably, ZnO-NPs demonstrated improved chemosensitivity by lowering the half-maximal inhibitory concentration (IC_{50}) of cisplatin in drug-resistant GC cell lines. ZnO-NPs mechanistically inhibited autophagy, as demonstrated by elevated p62 expression and decreased levels of Beclin-1 and LC3II/LC3I. ZnO-NPs were shown to reduce tumor growth in chemo resistant GC models in in vivo tests. These results imply that by blocking autophagy, ZnO-NPs could be used as a therapeutic treatment to help GC patients overcome chemoresistance (Miao et al., 2021). Because of their special optical and physicochemical characteristics, gold nanoparticles (AuNPs) are perfect for cancer treatment. In HCC cells, gold nanoparticles (GNPs) coupled with doxorubicin, capecitabine, and cisplatin improved drug retention, decreased drug efflux, and overcame chemoresistance. Increased apoptosis and decreased tumor cell proliferation resulted from their inhibition of P-gp-mediated resistance. The treatment efficacy was enhanced by the nanoparticle-drug complexes, which demonstrated more cytotoxicity than free medicines. This approach improves the results of treatment in HCC while successfully avoiding MDR (Tomuleasa et al., 2012). To combat MDR in breast cancer, a study created chondroitin sulfate-coated mesoporous silica nanoparticles (MSNs) co-loaded with quercetin (QC) and paclitaxel (PTX). By specifically targeting CD44 receptors, the nanoparticles improved uptake in MCF-7/ADR cells. Higher cytotoxicity was demonstrated by the co-loaded formulation, which caused G2/M cell cycle arrest and apoptosis. Quercetin increased paclitaxel sensitivity by inhibiting P-gp. The nanoparticles showed efficient tumor suppression and targeted tumor accumulation in vivo. A promising tactic to counteract MDR in breast cancer treatment is provided by this dual-drug nanoplatform (Liu et al., 2020). Because of their exceptional and distinctive qualities, inorganic mesoporous silica nanoparticles are the most recent additions to the field of nanoparticles. Mesoporous silica is a nanoporous SiO_2 material that can be functionalized for biomedical applications because of its numerous silanol (Si–OH) groups, high surface area (>1000 m^2/g), and highly organized pore structure (2–50 nm). It is useful for drug delivery, biosensing, and catalysis because of its stability, biocompatibility, and adjustable porosity (Vallet-Regí et al., 2017).

Novel drug targeting ferroptosis

Ferroptosis is a cell death mechanism distinct from apoptosis, necrosis, and autophagy. It is an intracellular process. Ferroptosis's unique mechanisms and its capacity to overcome resistance associated with traditional medicines make it a promising therapeutic approach for the treatment of cancer. Ferroptosis is an iron-dependent, oxidative stress-driven form of cell death that bypasses conventional resistance mechanisms, in contrast to apoptosis, which many cancer cells evade through genetic changes (Zhang et al., 2022). Furthermore, iron excess is frequently observed in cancer cells, particularly those with high metabolic activity, which increases their vulnerability to oxidative damage induced by ferroptosis. Polyunsaturated fatty acids (PUFAs) are abundant in the membranes of many tumors due to deregulated lipid metabolism, which makes them susceptible to peroxidation. By promoting lipid peroxidation and ROS accumulation, ferroptosis exploits this vulnerability to cause irreversible membrane damage and cell death (Piccolo et al., 2024; Sokol et al., 2024; Suda et al., 2024).

A formulation containing amorphous calcium carbonate (ACC) nanoparticles, which include a doxorubicin (DOX)-ferrous ion (Fe^{2+}) combination, was prepared to trigger ferroptosis. As a protective carrier, the ACC stabilizes the Fe^{2+} and prevents the medication from releasing too soon into the bloodstream. Once absorbed, the ACC carrier is broken down by the acidic environment in tumor cell lysosomes, releasing the DOX-Fe^{2+} complex. This complex decomposes into free DOX and Fe^{2+}, with DOX activating NADPH oxidase 4 (NOX4) to produce ROS. Lipid peroxidation and ferroptosis are induced by the ROS, especially hydrogen peroxide ($H2O_2$), reacting with $Fe2^+$ to create hydroxyl radicals via the Fenton reaction. Simultaneously, DOX induces apoptosis by inhibiting topoisomerase 2. Strong tumor cell killing is the result of this dual ferroptosis-apoptosis process (Suda et al., 2024).

Combination therapy

Several inhibitors targeting the PI3K/AKT/mTOR signaling transduction pathway have been evaluated in phase I or II clinical trials. These inhibitors could also be used in combination with chemo- or radiotherapy to restore the sensitivity of castration-resistant prostate cancer patients to traditional treatments (Su et al., 2024). For BRCA-mutated TNBC, poly (ADP-ribose) polymerase inhibitors (PARPis) have emerged as a promising targeted treatment, exploiting vulnerabilities in the HR repair (HRR) pathway. The efficacy of PARPis is often compromised by the development of resistance mechanisms, which include restoration of HRR, replication fork stabilization, reduced PARP1 trapping, and drug efflux. PARPi resistance is overcome through combination therapies, which include the integration of PARPis with chemotherapy, immunotherapy, antibody-drug conjugates, and PI3K/AKT pathway inhibitors, enhancing the therapeutic efficacy of PARPis by targeting multiple cancer progression pathways (Jain et al., 2025). Crizotinib, a tyrosine kinase inhibitor, enhanced the efficacy of PARP inhibitors in ovarian cancer cells and xenograft models by inducing autophagy. These results suggest a potential synergy between PARP inhibitors and crizotinib. This combination could be considered in patients to limit resistance to PARP inhibitors (Musa et al., 2024). Lovastatin sensitizes GBC cells to cisplatin-induced apoptosis and inhibits CHK1, CHK2, and H2AX activation during the DDR. It also leads to suppression of cholesterol biosynthesis, which inhibits GBC cell proliferation, possibly through attenuating the DNA repair process (Zhang et al., 2019). Metformin synergistically enhances the antitumor activity of cisplatin in GBC through PI3K/AKT/ERK signaling. Verapamil also induces the production of intracellular ROS, reduces GSH levels, and enhances the chemosensitivity of GBC cells to platinum drugs (Lai et al., 2023).

Targeting autophagy with natural compounds

For the management and treatment of a variety of human diseases, including cancer, by inhibiting autophagy pathways, natural products have historically been regarded as integral components of traditional medicine or folk medicine. It has been observed that the inhibition of autophagy using natural compounds can effectively enhance cancer cell death induced by diverse anticancer drugs through the specific targeting of various structures in the autophagy pathway, which include isolation membranes, autophagosomes, autolysosomes, and lysosomes (Al-Bari et al., 2021). Isoliquiritigenin (ISL) is a licorice flavonoid with antioxidant, antiinflammatory, and tumor-suppressing effects. ISL inhibits angiogenesis and breast cancer growth in a female nude mice model (Wang et al., 2021a). Resveratrol is a polyphenolic compound derived from red wine and red fruits with well-documented antitumor effects. Resveratrol treatment also increased the expression of p-AMPKα, which has been associated with its ability to induce apoptosis and repress autophagy through the inhibition of the mTOR signaling pathway. *Fucus vesiculosus* (brown seaweed) extracts appear to display anticancer effects in both ER-dependent and independent ways in HEC-1-B and RL95-2 endometrial cancer cell lines by competitively inhibiting estradiol (E2) binding to the ER (Nuñez-Olvera et al., 2019).

Personalized/precision medicine

The revolutionary approach of prescribing medicaments based on genetic makeup, molecular profile, lifestyle, and environmental factors is known as personalized medicine. The US National Research Council preferred the term "precision medicine" and regarded

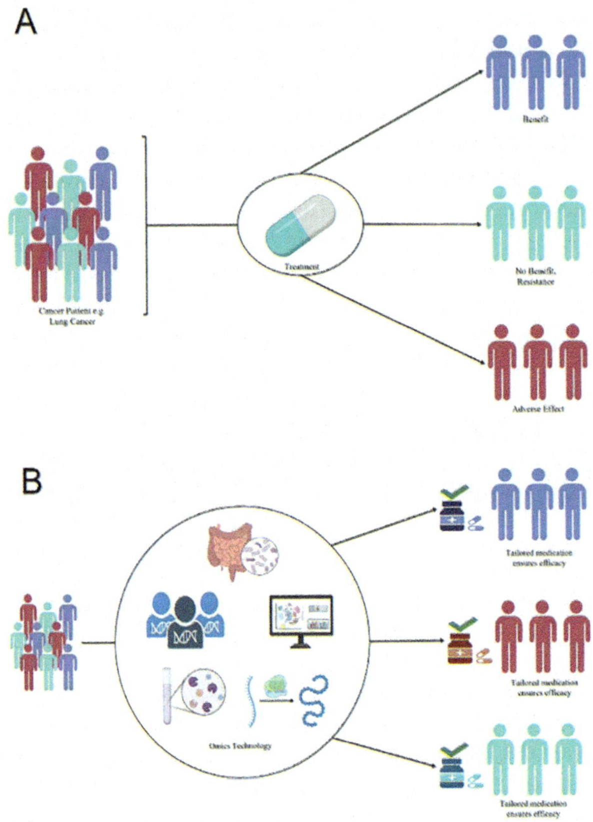

FIG. 5.4 (A) The idea of "one size fits all" is utilized in traditional medicine, which means that a single medication or specific class of medication is used to treat all individuals with a given illness. This results in therapeutic issues such as medication resistance, ineffectiveness, and severe drug reactions, among others. (B) Big data, artificial intelligence, machine learning, pharmacoomics, and omics are some of the methods used in personalized medicine to customize drugs for each patient based on all the variables that can affect drug response in a population. Genomics is the study of an organism's entire DNA collection, or genome (Guo et al., 2025). Transcriptomics, or the study of both coding and noncoding RNA (transcriptome), is used to investigate gene expression. Proteomics examines the entire protein (proteome) makeup of a cell, detecting alterations and interactions that are critical to biological functions. The metabolome is the total set of metabolites present in a biological system, such as a cell, tissue, or organism; metabolomics is the study of this collection. Precision and personalized medicine are revolutionizing cancer therapy by tailoring therapeutic approaches to each patient's particular characteristics, resulting in more targeted and effective treatments.

"personalized medicine" as an older term. Thus the terms "precision medicine" and "personalized medicine" are used interchangeably. The concept of personalized medicine was first developed in the field of oncology and later expanded into other areas of medicine. Traditional medicine uses the "one-size-fits-all" concept, which means a single drug or a particular class of drugs is used to treat all patients suffering from a disease. This leads to therapeutic problems such as adverse drug reactions, lack of efficacy, drug resistance, and so on. These variations in responses after drug administration could be due to factors such as gender, age, environment, lifestyle, occupation, ethnicity, genetic variations, concomitant medication, comorbidities, familial history of similar diseases, etc. Taking all these factors into consideration, the practice of personalized medicine uses tools such as omics, pharmacoomics, big data, artificial intelligence, and machine learning to tailor medications to individuals based on all factors that can lead to population variability in drug response (Naithani et al., 2021; Jafari et al., 2014).

Multiomics methods, such as genomics, transcriptomics, proteomics, epigenomics, microbiomics, and metabolomics, offer a comprehensive picture of a person's biological composition. The study of an organism's complete set of DNA (genome) is known as genomics (Hasin et al., 2017). To study gene expression, transcriptomics is applied, which is the study of both coding and noncoding RNA (transcriptome). Proteomics examines a cell's entire protein (proteome) composition, identifying changes and interactions that are essential to biological processes. The entire collection of metabolites found in a biological system, such as a cell, tissue, or organism, is referred to as the metabolome, and its analysis is called metabolomics. By adjusting therapeutic approaches to each patient's unique traits, personalized and precision medicine are transforming the treatment of cancer and producing more focused and efficient therapies (Hasanzad et al., 2021; Pammi et al., 2022). The general depiction of traditional and precision medicine is shown in Fig. 5.4 (Bray et al., 2024).

Precision medicine has greatly benefited from the discovery of biomarkers in malignancies, such as EGFR mutations, ALK gene fusions, and HER2 overexpression, thanks to developments in omics technologies (La Thangue and Kerr, 2011; Michelotti et al., 2022; Fang et al., 2015; Liu et al., 2017; Kerr and López-Ríos, 2016). Trastuzumab and pertuzumab are currently effective treatments for HER2-positive breast cancer, and osimertinib is beneficial for EGFR-mutated nonsmall cell lung cancer. ALK inhibitors like crizotinib and BRAF-targeted treatments like vemurafenib have also led to notable progress in treating ALK-rearranged lung cancer and BRAF-mutated melanoma (Musyuni et al., 2022). By targeting the molecular processes that cause cancer, these therapies offer a more potent substitute for conventional chemotherapy. Genomic sequencing, expression profiling, functional validation, correlations from clinical trials, and liquid biopsy methods have been used to identify biomarkers. By enabling tailored, biomarker-driven therapies, enhancing patient outcomes, and cutting down on unnecessary treatments, these findings have revolutionized oncology.

Conclusion

A comprehensive understanding of the different mechanisms of chemoresistance and the development of innovative treatment strategies are required for overcoming cancer resistance. Prioritizing the elucidation of the molecular basis of drug resistance and discovering advanced technologies to identify new targets is the urgent need of the hour. Combination therapies that address multiple resistance pathways should be developed, and personalized medicine approaches should be employed, as these would be critical in improving treatment outcomes.

Chemoresistance remains a major problem in cancer treatment as it reduces the efficacy of traditional chemotherapy and increases the risk of disease recurrence. More focused and efficient treatment approaches have become possible due to our growing understanding of the molecular mechanisms underlying chemoresistance, including drug efflux, altered drug metabolism, apoptosis evasion, DNA repair mechanisms, and interactions with the TME.

Recent developments in innovative drug formulations, such as nanoparticle-based delivery systems, have demonstrated promise in improving drug bioavailability, circumventing resistance mechanisms, and lowering systemic toxicity. In addition, the use of genetic and proteomic profiling to guide precision medicine offers a revolutionary way to customize cancer treatments according to patient profiles, enhancing therapeutic results while reducing side effects. In the future, developing cancer treatments will require a multidisciplinary strategy that incorporates innovative medication formulations, precision medicine based on biomarkers, and a better comprehension of tumor resistance pathways. Future studies can propel the creation of more potent, patient-centered cancer medicines by bridging the gap between preclinical discoveries and clinical applications, thereby increasing survival rates and quality of life for cancer patients throughout the globe.

References

Al-Bari, M.A.A., Ito, Y., Ahmed, S., Radwan, N., Ahmed, H.S., Eid, N., 2021. Targeting autophagy with natural products as a potential therapeutic approach for cancer. Int. J. Mol. Sci. 22 (18), 9807.

Ampolini, E.A., Jimenez-Sainz, J., Long, D.T., 2025. The development of ATM inhibitors in cancer therapy. Target Oncol. 20 (2), 281–297.

Baek, J.S., Kim, J.H., Park, J.S., Cho, C.W., 2015. Modification of paclitaxel-loaded solid lipid nanoparticles with 2-hydroxypropyl-β-cyclodextrin enhances absorption and reduces nephrotoxicity associated with intravenous injection. Int. J. Nanomed. 10, 5397–5405.

Balihodzic, A., Prinz, F., Dengler, M.A., Calin, G.A., Jost, P.J., Pichler, M., 2022. Non-coding RNAs and ferroptosis: potential implications for cancer therapy. Cell Death Differ. 29 (6), 1094–1106.

Boutayeb, A., 2010. The burden of communicable and non-communicable diseases in developing countries. In: Preedy, V.R., Watson, R.R. (Eds.), Handbook of Disease Burden and Quality of Life Measurement, Springer Science+Business Media LLC. pp. 531–546.

Bray, F., Laversanne, M., Sung, H., Ferlay, J., Siegel, R.L., Soerjomataram, I., et al., 2024. Global cancer statistics 2022: GLOBOCAN estimates of incidence and mortality worldwide for 36 cancers in 185 countries. CA Cancer J. Clin. 74 (3), 229–263.

Brunner, T.B., 2016. The rationale of combined radiotherapy and chemotherapy - Joint action of Castor and Pollux. Best Pract. Res. Clin. Gastroenterol. 30 (4), 515–528.

Bukowski, K., Kciuk, M., Kontek, R., 2020. Mechanisms of multidrug resistance in cancer chemotherapy. Int. J. Mol. Sci. 21 (9), 3233.

Chen, Q., Yang, Z., Liu, H., Man, J., Oladejo, A.O., Ibrahim, S., et al., 2024. Novel drug delivery systems: an important direction for drug innovation research and development. Pharmaceutics 16 (5), 674.

Christowitz, C., Davis, T., Isaacs, A., van Niekerk, G., Hattingh, S., Engelbrecht, A.M., 2019. Mechanisms of doxorubicin-induced drug resistance and drug resistant tumour growth in a murine breast tumour model. BMC Cancer 19 (1), 757.

Chu, X., Tian, W., Ning, J., Xiao, G., Zhou, Y., Wang, Z., Zhai, Z., Tanzhu, G., Yang, J., Zhou, R., 2024. Cancer stem cells: advances in knowledge and implications for cancer therapy. Signal Transduct. Target Ther. [Internet]. [cited 2025 Mar 20]. Available from: https://www.nature.com/articles/s41392-024-01851-y.

Criscitiello, C., Fumagalli, D., Saini, K.S., Loi, S., 2010. Tamoxifen in early-stage estrogen receptor-positive breast cancer: overview of clinical use and molecular biomarkers for patient selection. Oncol. Targets Ther. 4, 1–11.

Deepika, P.M., Salar, A., Salar, R.K., 2022. *In vitro* anticancer activity of curcumin loaded chitosan nanoparticles (CLCNPs) against Vero cells. Pharmacol. Res. – Mod. Chin. Med. 3, 100116.

Drayton, R.M., Dudziec, E., Peter, S., Bertz, S., Hartmann, A., Bryant, H.E., et al., 2014. Reduced expression of miRNA-27a modulates cisplatin resistance in bladder cancer by targeting the cystine/glutamate exchanger SLC7A11. Clin. Cancer Res. Off. J. Am. Assoc. Cancer Res. 20 (7), 1990–2000.

ER, S., JQ, W., DH, Y., XX, X., 2022. Paclitaxel resistance related to nuclear envelope structural sturdiness. Drug Resist. Update 65, 100881.

Fang, B., Mehran, R.J., Heymach, J.V., Swisher, S.G., 2015. Predictive biomarkers in precision medicine and drug development against lung cancer. Chin. J. Cancer 34 (3), 26.

Fongang, H., Mbaveng, A.T., Kuete, V., 2024. Chapter One - Cancer, global burden, and drug resistance. In: Kuete, V. (Ed.), Advances in Botanical Research [Internet]. (African Plant-based Products as a Source of Potent Drugs to Overcome Cancers and Their Chemoresistance Part 1. Cancer Chemoresistance, Screening Methods, and the Updated Cut-off Points for the Classification of Natural Cytotoxic Products), Vol. 111. Academic Press, pp. 1–51. [cited 2025 Mar 19]. Available from: https://www.sciencedirect.com/science/article/pii/S0065229623000812.

Gale, R.P., 2024. Chemotherapy and other systemic cancer treatments - cancer. MSD Manual Consumer Version [Internet]. [cited 2025 Mar 19]. Available from: https://www.msdmanuals.com/home/cancer/prevention-and-treatment-of-cancer/chemotherapy-and-other-systemic-cancer-treatments.

Gao, Z., Luan, X., Wang, X., Han, T., Li, X., Li, Z., et al., 2024. DNA damage response-related ncRNAs as regulators of therapy resistance in cancer. Front. Pharmacol. 15, 1390300.

Giglia-Mari, G., Zotter, A., Vermeulen, W., 2011. DNA damage response. Cold Spring Harb. Perspect. Biol. 3 (1), a000745.

Gottesman, M.M., 2002. Mechanisms of cancer drug resistance. Annu. Rev. Med. 53, 615–627.

Guo, Q., Zhou, Y., Xie, T., Yuan, Y., Li, H., Shi, W., et al., 2023. Tumor microenvironment of cancer stem cells: perspectives on cancer stem cell targeting. Genes Dis 11 (3), 101043.

Guo, W., Wang, Y., Wang, Z., YP, W., Zheng, H., 2016. Inhibiting autophagy increases epirubicin's cytotoxicity in breast cancer cells. Cancer Sci. 107 (11), 1610–1621.

Guo, X.J., Zhu, B.B., Li, J., Guo, P., Niu, Y.B., Shi, J.L., et al., 2025. Cholesterol metabolism in tumor immunity: mechanisms and therapeutic opportunities for cancer. Biochem. Pharmacol. 234, 116802.

Hasanzad, M., Sarhangi, N., Ehsani Chimeh, S., Ayati, N., Afzali, M., Khatami, F., et al., 2021. Precision medicine journey through omics approach. J. Diabetes Metab. Disord. 21 (1), 881–888.

Hasin, Y., Seldin, M., Lusis, A., 2017. Multi-omics approaches to disease. Genome Biol. Full Text [Internet]. [cited 2025 Mar 20]. Available from: https://genomebiology.biomedcentral.com/articles/10.1186/s13059-017-1215-1.

Herdiana, Y., Wathoni, N., Shamsuddin, S., Muchtaridi, M., 2022. Drug release study of the chitosan-based nanoparticles. Heliyon 8 (1), e08674.

Hientz, K., Mohr, A., Bhakta-Guha, D., Efferth, T., 2017. The role of p53 in cancer drug resistance and targeted chemotherapy. Oncotarget 8 (5), 8921–8946.

Jafari, S., Abdollahi, M., Saeidnia, S., 2014. Personalized medicine: a confluence of traditional and contemporary medicine. Altern. Ther. Health Med. 20 (5), 31–40.

Jain, A., Barge, A., Parris, C.N., 2025. Combination strategies with PARP inhibitors in BRCA-mutated triple-negative breast cancer: overcoming resistance mechanisms. Oncogene 44 (4), 193–207.

Jiang, M., Jia, K., Wang, L., Li, W., Chen, B., Liu, Y., et al., 2021. Alterations of DNA damage response pathway: biomarker and therapeutic strategy for cancer immunotherapy. Acta Pharm. Sin. B 11 (10), 2983–2994.

Jiang, W., Jin, W.L., Xu, A.M., 2024. Cholesterol metabolism in tumor microenvironment: cancer hallmarks and therapeutic opportunities. Int. J. Biol. Sci. 20 (6), 2044–2071.

Jiang, Y., Chen, M., Nie, H., Yuan, Y., 2019. PD-1 and PD-L1 in cancer immunotherapy: clinical implications and future considerations. Hum. Vaccines Immunother. 15 (5), 1111–1122.

Jurkovicova, D., Neophytou, C.M., Gašparović, A.Č., Gonçalves, A.C., 2022. DNA damage response in cancer therapy and resistance: challenges and opportunities. Int. J. Mol. Sci. 23 (23), 14672.

Kang, K.W., Chun, M.K., Kim, O., Subedi, R.K., Ahn, S.G., Yoon, J.H., et al., 2010. Doxorubicin-loaded solid lipid nanoparticles to overcome multidrug resistance in cancer therapy. Nanomedicine 6 (2), 210–213.

Kerr, K.M., López-Ríos, F., 2016. Precision medicine in NSCLC and pathology: how does *ALK* fit in the pathway? Ann. Oncol. 27, iii16–24.

Khan, S.U., Fatima, K., Aisha, S., Malik, F., 2024. Unveiling the mechanisms and challenges of cancer drug resistance. Cell Commun. Signal. 22 (1), 109.

Kim, S.W., young, K.H., Chi, D.W., Shim, J.H., Park, J.D., Lee, Y.H., et al., 2003. Reversal of P-glycoprotein-mediated multidrug resistance by ginsenoside Rg3. Biochem. Pharmacol. 65 (1), 75–82.

Koppula, P., Zhuang, L., Gan, B., 2021. Cystine transporter SLC7A11/xCT in cancer: ferroptosis, nutrient dependency, and cancer therapy. Protein Cell 12 (8), 599–620.

Kumar, S., Fairmichael, C., Longley, D.B., Turkington, R.C., 2020a. The Multiple Roles of the IAP Super-family in cancer. Pharmacol. Ther. 214, 107610.

Kumar, S., Gonzalez, E.A., Rameshwar, P., Etchegaray, J.P., 2020b. Non-coding RNAs as mediators of epigenetic changes in malignancies. Cancers 12 (12), 3657.

Kwon, J.E., Jang, Y., Yun, B.S., Kang, S., Kim, Y.H., Kim, B.G., et al., 2024. MET overexpression in ovarian cancer via CD24-induced downregulation of miR-181a: a signalling for cellular quiescence-like state and chemoresistance in ovarian CSCs. Cell Prolif. 57 (5), e13582.

La Thangue, N.B., Kerr, D.J., 2011. Predictive biomarkers: a paradigm shift towards personalized cancer medicine. Nat. Rev. Clin. Oncol. 8 (10), 587–596.

Lai, J., Yang, S., Lin, Z., Huang, W., Li, X., Li, R., et al., 2023. Update on chemoresistance mechanisms to first-line chemotherapy for gallbladder cancer and potential reversal strategies. Am. J. Clin. Oncol. 46 (4), 131–141.

Lee, J.H., Chae, J.W., Kim, J.K., Kim, H.J., Chung, J.Y., Kim, Y.H., 2015. Inhibition of cisplatin-resistance by RNA interference targeting metallothionein using reducible oligo-peptoplex. J. Control Release Off. J. Control Release Soc. 215, 82–90.

Lei, Z.N., Tian, Q., Teng, Q.X., Wurpel, J.N.D., Zeng, L., Pan, Y., et al., 2023. Understanding and targeting resistance mechanisms in cancer. MedComm 4 (3), e265.

Li, Y.Q., Yin, J.Y., Liu, Z.Q., Li, X.P., 2018. Copper efflux transporters ATP7A and ATP7B: novel biomarkers for platinum drug resistance and targets for therapy. IUBMB Life 70 (3), 183–191.

Lin, W., Wang, C., Liu, G., Bi, C., Wang, X., Zhou, Q., et al., 2020. SLC7A11/xCT in cancer: biological functions and therapeutic implications. Am. J. Cancer Res. 10 (10), 3106–3126.

Liu, B., Zhou, H., Tan, L., Siu, K.T.H., Guan, X.Y., 2024. Exploring treatment options in cancer: tumor treatment strategies. Signal Transduct. Target Ther. 9 (1), 1–44.

Liu, G., Yang, L., Chen, G., Xu, F., Yang, F., Yu, H., et al., 2021. A review on drug delivery system for tumor therapy. Front. Pharmacol. 12, 735446.

Liu, M., Fu, M., Yang, X., Jia, G., Shi, X., Ji, J., et al., 2020. Paclitaxel and quercetin co-loaded functional mesoporous silica nanoparticles overcoming multidrug resistance in breast cancer. Colloids Surf. B Biointerfaces 196, 111284.

Liu, P., Chen, G., Zhang, J., 2022. A review of liposomes as a drug delivery system: current status of approved products, regulatory environments, and future perspectives. Molecules 27 (4), 1372.

Liu, X., Wang, P., Zhang, C., Ma, Z., 2017. Epidermal growth factor receptor (EGFR): a rising star in the era of precision medicine of lung cancer. Oncotarget 8 (30), 50209–50220.

Ludeman, S.M., Gamcsik, M.P., 2002. Mechanisms of resistance against cyclophosphamide and ifosfamide: can they be overcome without sacrificing selectivity? In: Andersson, B., Murray, D. (Eds.), Clinically Relevant Resistance in Cancer Chemotherapy [Internet]. Springer US, Boston, MA, pp. 177–197. [cited 2025 Mar 19]. Available from: https://doi.org/10.1007/978-1-4615-1173-1_9.

Ma, X., Yu, H., 2006. Global burden of cancer. Yale J. Biol. Med. 79 (3–4), 85–94.

Makadia, H.K., Siegel, S.J., 2011. Poly lactic-co-glycolic acid (PLGA) as biodegradable controlled drug delivery carrier. Polymers 3 (3), 1377–1397.

Mansoori, B., Mohammadi, A., Davudian, S., Shirjang, S., Baradaran, B., 2017. The different mechanisms of cancer drug resistance: a brief review. Adv. Pharm. Bull. 7 (3), 339–348.

Mao, X., Lee, N.K., Saad, S.E., Fong, I.L., 2024. Clinical translation for targeting DNA damage repair in non-small cell lung cancer: a review. Transl. Lung Cancer Res. 13 (2), 375–397.

Mercogliano, M.F., Bruni, S., Mauro, F.L., Schillaci, R., 2023. Emerging targeted therapies for HER2-positive breast cancer. Cancers 15 (7), 1987.

Miao, Y.H., Mao, L.P., Cai, X.J., Mo, X.Y., Zhu, Q.Q., Yang, F.T., et al., 2021. Zinc oxide nanoparticles reduce the chemoresistance of gastric cancer by inhibiting autophagy. World J. Gastroenterol. 27 (25), 3851–3862.

Michelotti, A., de Scordilli, M., Bertoli, E., De Carlo, E., Del Conte, A., Bearz, A., 2022. NSCLC as the paradigm of precision medicine at its finest: the rise of new druggable molecular targets for advanced disease [Internet]. Int. J. Mol. Sci. [cited 2025 Mar 20]. Available from: https://www.mdpi.com/1422-0067/23/12/6748.

Mo, K., Kim, A., Choe, S., Shin, M., Yoon, H., 2023. Overview of solid lipid nanoparticles in breast cancer therapy. Pharmaceutics 15 (8), 2065.

Mukherjee, S., Ray, S., Thakur, R.S., 2009. Solid lipid nanoparticles: a modern formulation approach in drug delivery system. Indian J. Pharm. Sci. 71 (4), 349–358.

Musa, S., Amara, N., Selawi, A., Wang, J., Marchini, C., Agbarya, A., et al., 2024. Overcoming chemoresistance in cancer: the promise of crizotinib. Cancers 16 (13), 2479.

Musyuni, P., Bai, J., Sheikh, A., Vasanthan, K.S., Jain, G.K., Abourehab, M.A.S., et al., 2022. Precision medicine: ray of hope in overcoming cancer multidrug resistance. Drug Resist. Update 65, 100889.

Naithani, N., Sinha, S., Misra, P., Vasudevan, B., Sahu, R., 2021. Precision medicine: concept and tools. Med. J. Armed Forces India 77 (3), 249–257.

Niu, Q., Ye, S., Zhao, L., Qian, Y., Liu, F., 2024. The role of liver cancer stem cells in hepatocellular carcinoma metastasis. Cancer Biol. Ther. 25 (1), 2321768.

Nuñez-Olvera, S.I., Gallardo-Rincón, D., Puente-Rivera, J., Salinas-Vera, Y.M., Marchat, L.A., Morales-Villegas, R., et al., 2019. Autophagy machinery as a promising therapeutic target in endometrial cancer. Front. Oncol. 9, 1326.

O'Shaughnessy, J.A., 2003. Pegylated liposomal doxorubicin in the treatment of breast cancer. Clin. Breast Cancer 4 (5), 318–328.

Olusanya, T.O.B., Haj Ahmad, R.R., Ibegbu, D.M., Smith, J.R., Elkordy, A.A., 2018. Liposomal drug delivery systems and anticancer drugs. Molecules 23 (4), 907.

Osborne, M.P., Borgen, P.I., 1990. Role of mastectomy in breast cancer. Surg. Clin. North Am. 70 (5), 1023–1046.

Pammi, M., Aghaeepour, N., Neu, J., 2022. Multiomics, artificial intelligence, and precision medicine in perinatology. Pediatr. Res. [Internet]. [cited 2025 Mar 20]. Available from: https://www.nature.com/articles/s41390-022-02181-x.

Park, J.H., Zhuang, J., Li, J., Hwang, P.M., 2016. p53 as guardian of the mitochondrial genome. FEBS Lett 590 (7), 924–934.

Paul, R., Dorsey, J.F., Fan, Y., 2022. Cell plasticity, senescence, and quiescence in cancer stem cells: biological and therapeutic implications. Pharmacol. Ther. 231, 107985.

Piccolo, M., Ferraro, M.G., Iazzetti, F., Santamaria, R., Irace, C., 2024. Insight into iron, oxidative stress and ferroptosis: therapy targets for approaching anticancer strategies. Cancers 16 (6), 1220.

Pires, M.M., Emmert, D., Hrycyna, C.A., Chmielewski, J., 2009. Inhibition of P-glycoprotein-mediated paclitaxel resistance by reversibly linked quinine homodimers. Mol. Pharmacol. 75 (1), 92–100.

Rahman, M., Alam, K., Hafeez, A., Ilyas, R., Beg, S., 2021. Chapter 7 - Metallic nanoparticles in drug delivery and cancer treatment. In: Beg, S., Rahman, M., Choudhry, H., Souto, E.B., Ahmad, F.J. (Eds.), Nanoformulation Strategies for Cancer Treatment [Internet]. (Micro and Nano Technologies). Elsevier, pp. 107–119. [cited 2025 Mar 20]. Available from: https://www.sciencedirect.com/science/article/pii/B9780128210956000082.

Ramuta, T.Ž., Kreft, M.E., 2022. Mesenchymal stem/stromal cells may decrease success of cancer treatment by inducing resistance to chemotherapy in cancer cells. Cancers 14 (15), 3761.

Rathore, R., McCallum, J.E., Varghese, E., Florea, A.M., Büsselberg, D., 2017. Overcoming chemotherapy drug resistance by targeting inhibitors of apoptosis proteins (IAPs). Apoptosis Int. J. Program. Cell Death 22 (7), 898–919.

Renkonen-Sinisalo, L., Seppälä, T.T., Järvinen, H.J., Mecklin, J.P., 2017. Subtotal colectomy for colon cancer reduces the need for subsequent surgery in lynch syndrome. Dis. Colon Rectum 60 (8), 792–799.

Roszkowska, M., 2024. Multilevel mechanisms of cancer drug resistance. Int. J. Mol. Sci. 25 (22), 12402.

Safaei, S., Sajed, R., Shariftabrizi, A., Dorafshan, S., Saeednejad Zanjani, L., Dehghan Manshadi, M., et al., 2023. Tumor matrix stiffness provides fertile soil for cancer stem cells. Cancer Cell Int 23 (1), 143.

Sethy, C., Kundu, C.N., 2021. 5-Fluorouracil (5-FU) resistance and the new strategy to enhance the sensitivity against cancer: implication of DNA repair inhibition. Biomed. Pharmacother. 137, 111285.

Shen, D.W., Pouliot, L.M., Hall, M.D., Gottesman, M.M., 2012. Cisplatin resistance: a cellular self-defense mechanism resulting from multiple epigenetic and genetic changes. Pharmacol. Rev. 64 (3), 706–721.

Sokol, K.H., Lee, C.J., Rogers, T.J., Waldhart, A., Ellis, A.E., Madireddy, S., et al., 2024. Lipid availability influences ferroptosis sensitivity in cancer cells by regulating polyunsaturated fatty acid trafficking. Cell Chem. Biol. [Internet]. [cited 2025 Mar 20]; Available from: https://www.sciencedirect.com/science/article/pii/S2451945624004045.

Su, H., Huang, L., Zhou, J., Yang, G., 2024. Prostate cancer stem cells and their targeted therapies. Front. Cell Dev. Biol. 12, 1410102.

Su, Z., Liu, Y., Wang, L., Gu, W., 2025. Regulation of SLC7A11 as an unconventional checkpoint in tumorigenesis through ferroptosis. Genes Dis 12 (1), 101254.

Suda, A., Umaru, B.A., Yamamoto, Y., Shima, H., Saiki, Y., Pan, Y., et al., 2024. Polyunsaturated fatty acids-induced ferroptosis suppresses pancreatic cancer growth. Sci. Rep. 14 (1), 4409.

Tomuleasa, C., Soritau, O., Orza, A., Dudea, M., Petrushev, B., Mosteanu, O., et al., 2012. Gold nanoparticles conjugated with cisplatin/doxorubicin/capecitabine lower the chemoresistance of hepatocellular carcinoma-derived cancer cells. J. Gastrointest. Liver Dis. 21 (2), 187–196.

Tu, Z., Karnoub, A.E., 2022. Mesenchymal stem/stromal cells in breast cancer development and management. Semin. Cancer Biol. 86 (Pt 2), 81–92.

Tufail, M., Hu, J.J., Liang, J., He, C.Y., Wan, W.D., Huang, Y.Q., et al., 2024. Hallmarks of cancer resistance. iScience [Internet] 27 (6). [cited 2025 Mar 20]. Available from: https://www.cell.com/iscience/abstract/S2589-0042(24)01201-X.

Vaghari-Tabari, M., Hassanpour, P., Sadeghsoltani, F., Malakoti, F., Alemi, F., Qujeq, D., et al., 2022. CRISPR/Cas9 gene editing: a new approach for overcoming drug resistance in cancer. Cell. Mol. Biol. Lett. 27 (1), 49.

Vallet-Regí, M., Colilla, M., Izquierdo-Barba, I., Manzano, M., 2017. Mesoporous silica nanoparticles for drug delivery: current insights. Mol. J. Synth. Chem. Nat. Prod. Chem. 23 (1), 47.

Villalobos Gutiérrez, P.T., Muñoz Carrillo, J.L., Sandoval Salazar, C., Viveros Paredes, J.M., Gutiérrez Coronado, O., 2023. Functionalized metal nanoparticles in cancer therapy. Pharmaceutics 15 (7), 1932.

Vitto, V.A.M., Bianchin, S., Zolondick, A.A., Pellielo, G., Rimessi, A., Chianese, D., et al., 2022. Molecular mechanisms of autophagy in cancer development, progression, and therapy. Biomedicines 10 (7), 1596.

Wang, H., Fleishman, J.S., Cheng, S., Wang, W., Wu, F., Wang, Y., et al., 2024a. Epigenetic modification of ferroptosis by non-coding RNAs in cancer drug resistance. Mol. Cancer 23 (1), 177.

Wang, J., Qu, H., Jin, L., Zeng, W., Qin, L., Zhang, F., et al., 2011. Pegylated Phosphotidylethanolamine Inhibiting P-Glycoprotein Expression and Enhancing Retention of Doxorubicin in MCF7/ADR Cells. J. Pharm. Sci. 100 (6), 2267–2277.

Wang, K.L., Yu, Y.C., Hsia, S.M., 2021a. Perspectives on the role of isoliquiritigenin in cancer. Cancers 13 (1), 115.

Wang, L., Wang, X., Shen, L., Alrobaian, M., Panda, S.K., Almasmoum, H.A., et al., 2021b. Paclitaxel and naringenin-loaded solid lipid nanoparticles surface modified with cyclic peptides with improved tumor targeting ability in glioblastoma multiforme. Biomed. Pharmacother. 138, 111461.

Wang, L., Wang, X., Zhu, X., Zhong, L., Jiang, Q., Wang, Y., Tang, Q., Li, Q., Zhang, C., Wang, H., Zou, D., 2024b. Drug resistance in ovarian cancer: from mechanism to clinical trial. Mol. Cancer. Full Text [Internet]. [cited 2025 Mar 20]. Available from: https://molecular-cancer.biomedcentral.com/articles/10.1186/s12943-024-01967-3.

Wang, W., Zhou, M., Xu, Y., Peng, W., Zhang, S., Li, R., et al., 2021c. Resveratrol-loaded TPGS-resveratrol-solid lipid nanoparticles for multidrug-resistant therapy of breast cancer: in vivo and in vitro study. Front. Bioeng. Biotechnol. 9, 762489.

Wang, Y., Yu, H., Wang, S., Gai, C., Cui, X., Xu, Z., et al., 2021d. Targeted delivery of quercetin by nanoparticles based on chitosan sensitizing paclitaxel-resistant lung cancer cells to paclitaxel. Mater. Sci. Eng. C 119, 111442.

Wilczyński, B., Dąbrowska, A., Kulbacka, J., Baczyńska, D., 2024. Chemoresistance and the tumor microenvironment: the critical role of cell-cell communication. Cell Commun. Signal. CCS 22 (1), 486.

Wong, J.S.L., Kwok, G.G.W., Tang, V., BCW, L., Leung, R., Chiu, J., et al., 2021. Ipilimumab and nivolumab/pembrolizumab in advanced hepatocellular carcinoma refractory to prior immune checkpoint inhibitors. J. Immunother. Cancer 9 (2), e001945.

Wu, J.H., Cheng, T.C., Zhu, B., Gao, H.Y., Zheng, L., Chen, W.X., 2023. Identification of cuproptosis-related gene SLC31A1 and upstream LncRNA-miRNA regulatory axis in breast cancer. Sci. Rep. 13 (1), 18390.

Xie, Q., Liu, Y., Li, X., 2020. The interaction mechanism between autophagy and apoptosis in colon cancer. Transl. Oncol. 13 (12), 100871.

Yang, J., Zhou, Y., Xie, S., Wang, J., Li, Z., Chen, L., et al., 2021. Metformin induces Ferroptosis by inhibiting UFMylation of SLC7A11 in breast cancer. J. Exp. Clin. Cancer Res. CR 40 (1), 206.

Yoncheva, K., Tzankov, B., Yordanov, Y., Spassova, I., Kovacheva, D., Frosini, M., et al., 2020. Encapsulation of doxorubicin in chitosan-alginate nanoparticles improves its stability and cytotoxicity in resistant lymphoma L5178 MDR cells. J. Drug Deliv. Sci. Technol. 59, 101870.

Yoo, H.S., Lee, K.H., Oh, J.E., Park, T.G., 2000. In vitro and in vivo anti-tumor activities of nanoparticles based on doxorubicin–PLGA conjugates. J Control. Release 68 (3), 419–431.

Yu, X., Li, W., Liu, H., Deng, Q., Wang, X., Hu, H., et al., 2020. Ubiquitination of the DNA-damage checkpoint kinase CHK1 by TRAF4 is required for CHK1 activation. J. Hematol. Oncol. 13 (1), 40.

Yu, X., Zhao, H., Wang, R., Chen, Y., Ouyang, X., Li, W., et al., 2024. Cancer epigenetics: from laboratory studies and clinical trials to precision medicine. Cell Death Discov. 10 (1), 28.

Zhang, C., Liu, X., Jin, S., Chen, Y., Guo, R., 2022. Ferroptosis in cancer therapy: a novel approach to reversing drug resistance. Mol. Cancer 21 (1), 47.

Zhang, Y., Huang, F., Luo, Q., Wu, X., Liu, Z., Chen, H., et al., 2018. Inhibition of XIAP increases carboplatin sensitivity in ovarian cancer. Oncol. Targets Ther. 11, 8751–8759.

Zhang, Y., Liu, Y., Duan, J., Wang, H., Zhang, Y., Qiao, K., et al., 2019. Cholesterol depletion sensitizes gallbladder cancer to cisplatin by impairing DNA damage response. Cell Cycle (Georget. Tex.) 18 (23), 3337–3350.

Zhu, Y., Wang, A., Zhang, S., Kim, J., Xia, J., Zhang, F., et al., 2023. Paclitaxel-loaded ginsenoside Rg3 liposomes for drug-resistant cancer therapy by dual targeting of the tumor microenvironment and cancer cells. J. Adv. Res 49, 159–173.

Zou, D., Xin, X., Xu, Y., Xu, H., Huang, L., Xu, T., 2024. Improving the efficacy of immunotherapy for colorectal cancer: targeting tumor microenvironment-associated immunosuppressive cells. Heliyon 10 (16), e36446.

CHAPTER

6

Novel drug delivery systems for cancer therapy

Santenna Chenchula[a], *Arumilli Swetha*[b], *Anitha Kuttiappan*[c], *Phulen Sharma*[d], *Sadiyaakhtar S. Kazi*[e], and *Madhavrao Chavan*[e,f]

[a]Department of Pharmacology, AIIMS, Bhopal, Madhya Pradesh, India
[b]Department of Pharmaceutics, Aditya College of Pharmacy, Surampalem, Andhra Pradesh, India
[c]Department of Pharmacology, School of Pharmacy and Technology Management (SPTM), SVKM's NMIMS Deemed-to-University, Shirpur, India
[d]Department of Pharmacology, AIIMS, Guwahati, Assam, India
[e]Department of Pharmacology, Government Medical College, Chandrapur, Maharashtra, India
[f]Department of Pharmacology, AIIMS, Mangalagiri, Andhra Pradesh, India

Novel drug delivery systems in cancer management

Novel drug delivery systems (NDDSs) refer to cancer treatment strategies focusing on creating cutting-edge ways to deliver therapeutic chemicals to specific bodily locations (George, 2023). By optimizing medication delivery, reducing side effects, and offering targeted therapy, these systems aim to increase the effectiveness and safety of cancer treatments. By precisely controlling drug release, NDDSs help ensure that the therapeutic agent reaches the tumor site at optimal concentrations, thus maximizing its effectiveness while minimizing systemic exposure.

The scope of NDDSs in cancer management encompasses various approaches and tools that fall into many general categories (Salahshoori et al., 2024). Drug delivery systems based on nanotechnology include encapsulating and delivering anticancer medications using liposomes, nanoparticles (NPs), and nanoencapsulated drugs. By altering their surface features with ligands that bind to receptors overexpressed on cancer cells, these NPs can be tailored to target particular cell types or tissues, such as tumor cells.

https://doi.org/10.1016/B978-0-443-38307-6.00003-X

Polymeric drug delivery systems: These systems encapsulate anticancer medications and regulate their release using biocompatible polymers. Drugs can be released from polymeric systems in response to specific stimuli, including temperature or pH changes, or they can be released gradually over time to provide sustained therapeutic doses (Malhi and Sharma, 2022).

Microparticle-based drug delivery systems: Like NPs, microparticles are small particles that can be loaded with anticancer drugs and directed toward particular tissues or cells. When combined with other NDDSs, microparticles can be engineered to achieve controlled release of the medication.

Implantable drug delivery systems: In these systems, devices that release anticancer medications directly into the tumor or surrounding tissue, such as drug-eluting stents or microchips, are implanted. Tumors can be treated locally with implantable systems, potentially delivering prolonged, continuous medication.

Targeted drug delivery systems: This strategy uses ligands, antibodies, or peptides targeting tumor microenvironments or cancer cells. These targeting moieties can be conjugated to drugs or drug carriers to improve their specificity and reduce off-target effects (George, 2023; Vasir et al., 2005).

Gene delivery systems: NDDS can transfer genetic material, such as small interfering RNA (siRNA) or microRNA, to alter gene expression in cancer cells, in addition to traditional chemical medicines. Synergistic effects can be obtained by combining gene delivery devices with other NDDS techniques.

Local drug delivery systems: These methods involve administering anticancer medications directly to the tumor location, avoiding the bloodstream and minimizing adverse effects on the system. Examples include intravesical instillation for bladder cancer, intratumoral injections, and intraarterial chemotherapy for liver cancer.

Combination drug delivery systems: In addition, NDDSs can be utilized to deliver different medications sequentially or simultaneously, resulting in synergistic or complementary effects. Drug delivery systems that include multiple drugs can be engineered to target distinct signaling pathways and circumvent resistance mechanisms (Hu et al., 2010).

The development and application of NDDSs in cancer management have the potential to completely transform cancer treatment by improving therapeutic outcomes and patient quality of life, acknowledging that the effective transition of NDDSs from preclinical research to clinical settings necessitates a thorough assessment of their safety, effectiveness, and pharmacokinetics (Xin et al., 2017; Chaturvedi et al., 2023). In addition, the complexity and heterogeneity of cancer pose significant challenges for the design and optimization of NDDSs, highlighting the need for interdisciplinary collaboration between researchers, clinicians, and industry partners (Fig. 6.1).

Fundamentals of NDDSs in cancer management

At its core, NDDSs involve the development of innovative delivery methods to transport therapeutic agents to target sites within the body, ensuring that they reach tumor cells at optimal concentrations (Hong et al., 2023). This is particularly important in cancer management, where tumors' inherent heterogeneity and complexity can pose challenges to effective treatment (Yezdani et al., 2018). By utilizing NDDSs, drugs can be encapsulated in various

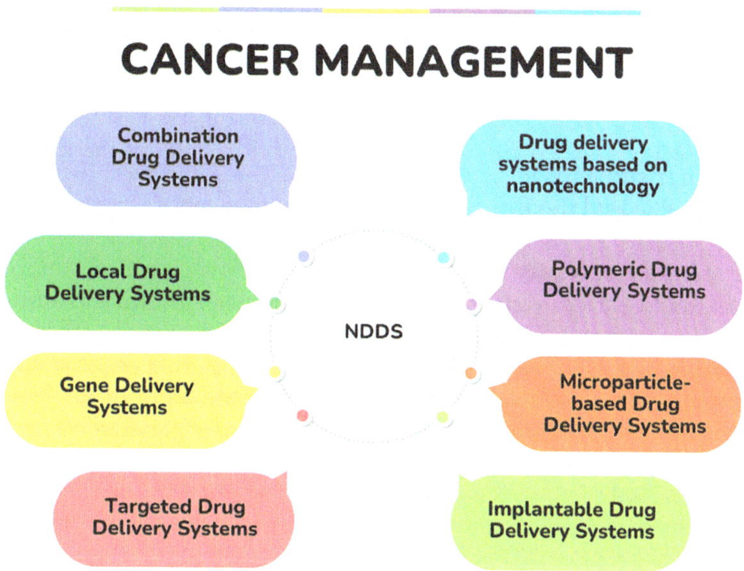

FIG. 6.1 Development and applications of NDDSs in cancer management.

nano- or microparticles, liposomes, or polymeric carriers, allowing for controlled and sustained drug release. This targeted delivery reduces the exposure of healthy tissues to the drug, thereby minimizing side effects. Moreover, NDDSs can be designed to exploit specific biological characteristics of cancer cells, such as overexpressed receptors or unique microenvironments, thus enhancing drug uptake by tumor cells while sparing normal cells (Raj et al., 2021). In addition, NDDSs can address challenges associated with conventional chemotherapy, such as drug resistance, by offering synergistic combinations of drugs or modalities and circumventing drug efflux mechanisms. In summary, the fundamentals of NDDSs in cancer management revolve around improving drug delivery, optimizing therapeutic efficacy, and enhancing patient outcomes while minimizing toxicity.

Evolution of NDDSs in cancer treatment

NDDSs have improved patient outcomes and quality of life, and their development in cancer treatment represents a significant advancement in oncology. The development of nanotechnology, materials science, and biotechnology has resulted in a notable metamorphosis of NDDSs over the last few decades. In the early stages of development, NDDSs primarily focused on encapsulating chemotherapeutic agents in liposomes and NPs to improve drug solubility, bioavailability, and pharmacokinetics, and reduce systemic toxicity. These early systems laid the foundation for more sophisticated NDDS approaches that have since been developed (Nilesh et al., 2019).

A significant turning point in the development of NDDSs was the advent of tailored drug delivery devices. The treatment efficacy of these systems is increased by minimizing off-target effects and focusing on cancer cells or tumor microenvironments individually. Drug carriers are coupled to targeting moieties, chemicals, peptides, or antibodies that bind to antigens or receptors overexpressed on cancer cells. By enabling greater medication concentrations at the tumor site with less harm to healthy tissues, targeted drug delivery devices have significantly increased the precision and selectivity of cancer treatments (Puccetti et al., 2024).

Developing stimuli-responsive drug delivery devices is a significant accomplishment of NDDSs. These systems are designed to react to particular stimuli, including variations in temperature, pH, or enzyme activity, to release the medicine at the intended location. To provide tailored therapy, pH-sensitive liposomes, for instance, release their contents in response to the acidic environment of tumors. In addition to providing exact control over drug release kinetics and enabling sustained or on-demand drug delivery, stimuli-responsive drug delivery systems also hold promise for circumventing drug resistance mechanisms (Rahim et al., 2021).

Gene delivery systems represent a revolutionary method for cancer treatment, developed in recent years due to the merging of genetic engineering with nanotechnology. These systems involve the transfer of genetic material to cancer cells to modify the expression of specific genes, such as short-interfering RNAs (siRNAs) or microRNAs. Gene delivery systems can target specific signaling pathways or circumvent medication resistance. They can also be used with other NDDSs to provide beneficial side effects (Hani et al., 2020).

A further factor influencing the development of NDDSs in cancer treatment is the introduction of immunotherapy. Some malignancies respond favorably to immunotherapeutic treatments, such as monoclonal antibodies and checkpoint inhibitors (Salahshoori et al., 2024). Through improved distribution, NDDSs can decrease systemic side effects and increase the effectiveness of these medicines. By altering the tumor microenvironment to strengthen the antitumor immune response, NPs can encapsulate immunotherapeutic drugs and transport them to the tumor site (George, 2023).

In terms of cancer treatment, NDDSs have a bright future. NDDSs integrate emerging technologies such as machine learning and artificial intelligence to develop more individualized and accurate cancer treatments. Complex drug delivery systems with exact control over drug release kinetics and spatial dispersion can now be created because of advancements in biomaterials and 3D printing. NDDSs will become increasingly crucial in creating novel and potent cancer treatments as our knowledge of cancer biology advances (Emerich, 2005).

Targeted drug delivery systems of NDDSs in cancer management

Several targeted drug delivery strategies are used by NDDSs in cancer treatment to increase the therapeutic efficacy of anticancer medications while lowering systemic toxicity (Di Stefano, 2023). An example of such a system is the enhanced permeability and retention (EPR) effect of NPs, which can be engineered to concentrate preferentially in tumors. Engineering NPs can increase the specificity of medication delivery by including tumor-specific ligands, such as peptides or antibodies, that bind to and recognize receptors overexpressed on cancer cells. Another NDDS platform is liposomes, lipid-based vesicles that contain medications and can have targeting ligands added to their surface. This allows them to engage in targeted interactions with cancer cells, promoting medication release within the cell and reducing drug exposure to healthy tissues (Puccetti et al., 2024). Furthermore, polymeric NDDSs are biocompatible polymers that can be modified for targeted delivery, and controlled drug release polymers can be coupled to targeting moieties or engineered to react to stimuli unique to tumors, including pH or temperature variations, to release drugs at specific sites. In addition, monoclonal antibodies that target tumor-specific antigens are used in antibody–drug conjugates (ADCs), which deliver a lethal payload specifically to cancer cells while sparing healthy cells. Finally, aptamer-based NDDSs use tiny, single-stranded DNA or RNA molecules that attach to particular cancer cell surface antigens. It is possible to conjugate these aptamers to drug carriers, allowing for the precise and targeted delivery of medication to cancer cells. These NDDS-targeted drug delivery systems show great potential for managing cancer by improving therapeutic efficacy and lowering adverse effects.

Advances in drug delivery systems

The field of pharmacotherapy has undergone tremendous changes due to advancements in drug delivery technologies, particularly in the treatment of complicated disorders such as cancer. The development of nanotechnology-based delivery systems, which take advantage of the special qualities of nanoscale materials to enhance medication solubility, stability, and targeting, is one noteworthy breakthrough. Accurate and effective drug administration can be achieved in target tissues, including tumor locations, using polymeric micelles, liposomes, and NPs that can encapsulate medications and cross physiological barriers. Another innovation is the creation of stimuli-responsive drug delivery systems, which can release medications in response to particular triggers such as temperature or pH changes or changes in enzyme activity (Puccetti et al., 2024). This minimizes adverse effects and maximizes therapeutic results by enabling the controlled delivery of medications in a spatially and temporally coordinated manner. Furthermore, the possibility of using gene therapy to cure cancer and genetic abnormalities has been made possible by advancements in gene delivery technologies. Viral and nonviral vectors are constructed to transfer genetic material—such as DNA or RNA—to target cells to change the expression of specific genes or fix genetic defects. In addition, the development of immunoliposomes, immune checkpoint inhibitors, and chimeric antigen receptor (CAR) T-cell treatments has revolutionized cancer treatment by integrating immunotherapeutic agents with drug delivery systems. These developments improve therapeutic efficacy and decrease side effects by enabling the exact control of the immune response against cancer cells. In conclusion, new developments in drug delivery technologies have advanced pharmacotherapy and opened up new avenues for focused and customized treatment for a variety of illnesses.

Liposomal drug delivery systems

Liposomal drug delivery systems are cutting-edge methods of drug administration that offer various benefits for treating and managing cancer (Kumar et al., 2021; Liu et al., 2021).

Liposomes are synthesized vesicles made of phospholipid bilayers that can encase both hydrophilic and hydrophobic medications. One of the primary benefits of liposomal delivery is its ability to shield the medication from deterioration and quick elimination, thereby lengthening the time it spends in the bloodstream (Salahshoori et al., 2024). By adding targeting ligands to their surface, liposomes can bind to particular cells or tissues, increasing drug bioavailability at the intended location (Sharma and Sharma, 1997). Moreover, liposomes allow controlled and sustained drug release, as they can be engineered to release the medication in response to particular stimuli such as temperature or pH changes or enzymatic destruction. The preservation of therapeutic concentrations at the target site not only decreases systemic toxicity but also improves drug efficacy (Kuna et al., 2024). Liposomal drug delivery systems have demonstrated encouraging outcomes in cancer treatment by enhancing the therapeutic efficacy of chemotherapeutic agents, minimizing side effects, and overcoming drug resistance. Liposomal formulations of anticancer medications such as doxorubicin and paclitaxel have been authorized for clinical use, demonstrating liposomal drug delivery systems' clinical applicability and promise in managing cancer.

Polymeric NPs

Among the medication delivery methods most often researched and used in contemporary pharmacology are polymeric NPs (Garg et al., 2018; Khizar et al., 2023; Gagliardi et al., 2021). Poly(lactic-co-glycolic acid) (PLGA) and polyethylene glycol, two biocompatible polymers, PLGA being additionally biodegradable, designed to encapsulate and shield medications, peptides, or genes, make up these NPs. The capacity to deliver prolonged and regulated drug release is one of the main benefits of polymeric NPs (Zielińska et al., 2020). Reducing the frequency of medication administration and maintaining therapeutic concentrations can be achieved by controlling the degradation of these polymers to release the drug over a longer period (Malhi and Sharma, 2022). To increase the specificity of polymeric NPs for target cells or tissues, they can be surface-modified with targeting ligands such as peptides or antibodies. This enhances drug delivery efficiency by allowing NPs to overcome physiological barriers and accumulate at the intended site. Moreover, polymeric NPs can be engineered to react to particular stimuli, such as temperature or pH changes, or enzymatic degradation, allowing for initiated drug release. This not only enhances drug efficacy but also minimizes systemic toxicity (Hamid and Manzoor, 2020). In cancer management, polymeric NPs have shown promising results in delivering chemotherapeutic agents, such as doxorubicin or paclitaxel, to tumor sites while reducing side effects. Moreover, these NPs can be engineered to encapsulate multiple drugs, offering synergistic effects or overcoming drug resistance. When combined, polymeric NPs offer a flexible and efficient medication delivery system that could transform personalized therapy for various illnesses.

Dendrimers and micelles

The special qualities and prospective uses of dendrimers and micelles, two advanced drug delivery methods, have attracted much interest in the field of pharmacology. Dendrimers are described as three-dimensional, highly branched macromolecules with monodisperse size distributions and clearly defined structures (Malhi and Sharma, 2022). They are excellent candidates for delivering therapeutic drugs because of their dendritic architecture, which provides precise control over drug loading and release kinetics (Svenson, 2009; Mintzer and Grinstaff, 2011). To improve drug delivery efficiency and reduce off-target effects, dendrimers can be surface-functionalized with targeting ligands, such as antibodies or peptides, to increase their selectivity for target cells or tissues (Khizar et al., 2023). Furthermore, dendrimers offer versatility in drug loading since they can be engineered to encapsulate hydrophobic pharmaceuticals within their internal cavities or coupled to hydrophilic drugs via covalent connections. Moreover, the abundance of functional groups on the surface of dendrimers makes it easy to modify them using imaging agents or diagnostic probes, opening up new possibilities for theranostic uses (Anitha et al., 2024).

Conversely, amphiphilic chemicals spontaneously combine into self-assembling structures called micelles in aqueous environments. Due to these structures' hydrophobic core and hydrophilic shell, hydrophobic medications can be dissolved inside the core and delivered to target cells (Garg et al., 2018). Controlled medication release and improved tumor penetration are made possible by modifying micelles' size and surface characteristics. Furthermore, to increase the selectivity and effectiveness of micelles, they can be functionalized with stimuli-responsive moieties or targeting ligands. In addition, because micelles are biocompatible and biodegradable, they can be used for sustained medication release, which minimizes the need for frequent administration (Emerich, 2005; Di Stefano, 2023).

Both dendrimers and micelles have demonstrated significant potential in cancer therapy, particularly in the delivery of chemotherapeutic agents, nucleic acids, and imaging agents. They are appealing candidates for personalized and targeted therapy because of their capacity to increase drug solubility, stability, and targeting while reducing systemic toxicity. Despite these promising results, further research is needed to address challenges such as scale-up production, regulatory approval, and clinical translation. Nonetheless, the versatile properties of dendrimers and micelles hold promise for revolutionizing drug delivery and cancer management (Nie et al., 2007; Madaan et al., 2014).

Quantum dots and nanoscale drug delivery systems

Quantum dots (QDs) are semiconductor nanocrystals with distinct optical and electrical characteristics, rendering them potentially useful instruments for medication delivery and cancer treatment (Marwadi et al., 2022). They can be utilized to track and label medication delivery systems in vivo because of their small size and variable emission characteristics (Puccetti et al., 2024). Medication delivery systems can be optimized with the help of this real-time visualization of medication release and distribution. Furthermore, QDs can be functionalized with targeting ligands, such as peptides or antibodies, on their surface to direct them toward particular tissues or cells. This reduces off-target effects and improves therapeutic outcomes by increasing the specificity and efficacy of drug delivery systems (Kuna et al., 2024).

The purpose of nanoscale drug delivery systems, which include liposomes, micelles, and NPs, is to encapsulate and transport therapeutic agents to specific tissues or cells (Brigger, Dubernet and Couvreur, 2002). By adjusting these systems' size, shape, and surface characteristics, drug release can be regulated and directed toward particular locations, such as tumors (Malhi and Sharma, 2022). Furthermore, it is possible to induce drug release using nanoscale drug delivery devices by engineering them to react to particular stimuli, such as temperature or pH changes or enzyme destruction. This promotes sustained and regulated medication release, enhancing therapeutic efficacy and reducing systemic toxicity.

There are various benefits to integrating QDs with nanoscale medication delivery devices. First, QDs have the potential to function as Nano sensors, enabling real-time tracking of medication release and dispersal. Second, the specificity and effectiveness of drug delivery systems are improved by their capacity to target particular tissues or cells. Third, the drug distribution mechanism can be observed noninvasively due to the application of QDs in imaging. Fourth, multimodal imaging of drug delivery systems can be achieved by combining QDs with other imaging modalities, such as computed tomography or magnetic resonance imaging. Overall, targeted and personalized cancer therapy advancement appears to be significantly enhanced by incorporating QDs into nanoscale drug delivery systems.

Biodegradable polymers and hydrogels

Currently, available drug delivery methods include both hydrogels and biodegradable polymers, which provide distinct benefits in terms of controlled drug release and targeted therapy (Kuna et al., 2024). Because they are biocompatible and break down into harmless byproducts that the body can metabolize and excrete, biodegradable polymers such as PLGA and poly(lactic acid) are frequently used in drug delivery systems. The regulated and prolonged release of therapeutic substances over time can be achieved by forming these polymers into NPs, microparticles, or implants. Personalized and targeted medication delivery is also possible since the degradation rates of biodegradable polymers can be adjusted to meet the required release profile. Furthermore, targeting ligands, such as peptides or antibodies, can be added to the surface of biodegradable polymers to improve their specificity for target cells or tissues. This will minimize off-target effects and increase the effectiveness of drug delivery.

On the other hand, hydrogels are networks of hydrophilic polymers arranged in three dimensions that can absorb substantial volumes of biological fluids or water (Puccetti et al., 2024; Snehal et al., 2021). Hydrogels are a good choice for drug delivery applications due to their high-water content and mechanical characteristics, resembling soft tissues, especially in healing burns, wounds, and tissue engineering (Salahshoori et al., 2024). Because of the porous structure of hydrogels, medicines, peptides, or growth factors can be incorporated into them and released in a regulated and sustained manner (Buwalda et al., 2017). Hydrogels can also be designed to react to particular stimuli, such as temperature changes, pH shifts, or enzymatic degradation, which enables triggered drug release. By doing so, systemic toxicity is reduced in addition to enhancing medication efficacy. Hydrogels can also be functionalized with targeting ligands, such as peptides or antibodies, to increase their specificity and effectiveness and facilitate targeted medication administration to particular tissues or cells (Ho et al., 2022). In summary, biodegradable polymers and hydrogels have a number of benefits for drug delivery systems, such as minimized systemic toxicity, targeted therapy, and regulated and sustained drug release. Among other uses, their biocompatibility, adjustable characteristics, and capacity to react to particular stimuli make them attractive options for targeted and customized cancer treatment (Fig. 6.2).

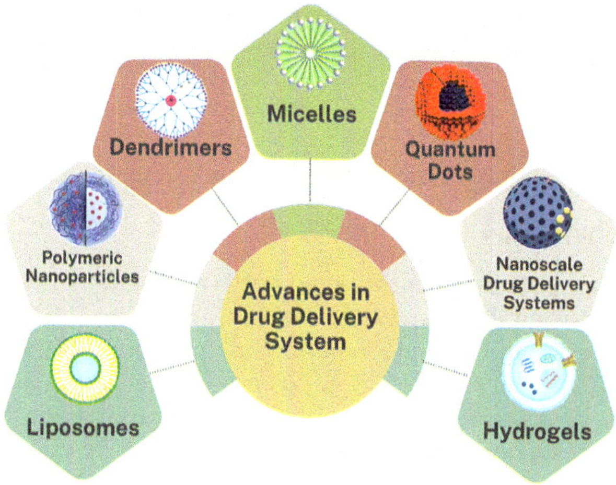

FIG. 6.2 Polymeric micelles, liposomes, and nanoparticles engineered to encapsulate medications enable precise and efficient delivery of drugs to target tissues, including tumors, by traversing physiological barriers effectively.

NDDS-targeted therapies in chemotherapy

Targeted treatments for cancer using NDDSs aim to maximize therapeutic specificity and efficacy while reducing systemic toxicity, marking a paradigm shift in oncology. These cutting-edge methods deliver therapeutic compounds directly to cancer cells or tumor microenvironments by utilizing precision medicine, biomaterial science, and advanced nanotechnology (Kuna et al., 2024).

The encapsulation and delivery of chemotherapeutic medications using liposomes or NPs are noteworthy examples of targeted therapeutics (Montané et al., 2020). By focusing on specific cancer cell receptors or tumor-associated indicators, these NDDS platforms can efficiently increase drug uptake by cancer cells while decreasing exposure to healthy tissues. Furthermore, drug resistance can be circumvented, and treatment effects can be enhanced by the regulated and sustained drug release offered by NPs and liposomes (Malhi and Sharma, 2022).

Targeted therapy has further potential through the use of gene delivery methods (Sun et al., 2023). Using NDDSs, genetic materials, such as microRNAs or small interfering RNAs (siRNAs), can be delivered to cancer cells to alter their gene expression. Addressing drug resistance or focusing on certain signaling pathways may improve the efficacy of cancer therapies (Gu et al., 2020). Moreover, incorporating immunotherapeutic drugs into NDDSs has transformed how cancer is treated. CAR T-cell treatments, immune checkpoint inhibitors, and monoclonal antibodies can all be encapsulated in liposomes or NPs to increase their specificity and effectiveness. Numerous malignancies, including metastatic melanoma and nonsmall cell lung cancer, respond very well to these targeted immunotherapies.

Personalized NDDSs have also made it possible to customize treatment plans according to a single patient's genetic profile, tumor features, and immunological response. Clinical professionals can improve drug selection and dosing, thereby enhancing patient outcomes, by combining NDDSs with biomarkers and molecular diagnostics (Fornaguera and García-Celma, 2017).

In NDDS-targeted treatments for cancer, treatment strategies are comprehensive approaches that use precision medicine and state-of-the-art technology. With better therapeutic results, fewer side effects, and an improved quality of life for patients, these cutting-edge approaches have significant potential to overcome the difficulties associated with traditional cancer treatments (Verma et al., 2017).

Nanocarriers for chemotherapy

The use of liposomes, polymeric micelles, and NPs as chemotherapeutic nanocarriers has shown promise in increasing the effectiveness of cancer treatments. Enhanced tumor targeting, regulated drug release, and extended circulation durations are just a few benefits offered by these nanocarriers (Vaibhav et al., 2020). Liposomes, which encapsulate hydrophilic pharmaceuticals in their aqueous core or hydrophobic medications in their lipid bilayer, represent one of the original types of nanocarriers. By utilizing the EPR effect or conjugation with targeting ligands, liposomes can selectively target cancer cells (Edis et al., 2021). Another type of nanocarrier that can solubilize hydrophobic medications within its core and stabilize them in aqueous environments is polymeric micelles (Edis et al., 2021). They have the ability to boost drug accumulation in tumors, extend drug circulation periods, and enhance drug stability. In conclusion, the special qualities of both inorganic and polymer-based NPs qualify them for use in chemotherapy (Begines et al., 2020). Polymer-based NPs can be tailored to possess certain drug release patterns and surface modifications, whereas inorganic NPs can offer several functions, including medication administration and imaging. In summary, using nanocarriers for chemotherapy is a viable way to optimize drug distribution, boost effectiveness, and lessen side effects while treating cancer.

Combination therapy and synergistic drug delivery

To enhance the results of cancer treatment, combination therapy—which entails the concurrent or sequential use of several medications or treatment modalities—has become prominent. With the goal of boosting therapeutic efficacy and lowering the risk of drug resistance, this method targets numerous pathways or mechanisms implicated in the progression of cancer. Combining combination therapy with innovative drug delivery systems (NDDSs) can be especially beneficial because it makes it possible to deliver therapeutic drugs to tumor locations precisely and specifically (Hu et al., 2010; Kydd et al., 2017).

Combining immunotherapy and chemotherapy to treat cancer is one example of combination therapy. Immunotherapy, which uses the body's immune system to target cancer cells, can be used in conjunction with chemotherapy, which uses cytotoxic medications to destroy cancer cells. Chemotherapy medications can be administered directly to tumor locations via NDDSs, while immunotherapy treatments, including monoclonal antibodies or checkpoint inhibitors, can be supplied either systemically or directly to the tumor microenvironment. Compared with monotherapy, this combined strategy has demonstrated improved response rates and survival outcomes in clinical trials.

The use of targeted therapy and NDDSs for cancer treatment is another form of combination therapy. The use of medications that selectively target cancer cells or tumor-associated chemicals, including growth factor receptors or angiogenesis inhibitors, is known as targeted therapy. These targeted therapeutic drugs can be more precisely and effectively delivered to tumor locations with the assistance of NDDSs. Targeted therapy can also be used in conjunction with other forms of treatment, including radiation or chemotherapy, to provide beneficial results. For instance, combining targeted therapy with chemotherapy can increase cancer cell susceptibility to drugs, improving the course of treatment. Combination therapy and synergistic NDDSs for cancer treatment represent promising approaches for improving therapeutic efficacy, reducing drug resistance, and minimizing side effects. These innovative strategies leverage the advantages of targeted therapy, immunotherapy, and drug delivery systems, offering new creative approaches for personalized and effective cancer treatment. However, combination therapy requires careful consideration of drug interactions, dose optimization, and patient-specific factors to achieve the desired therapeutic outcomes (Kydd et al., 2017).

Clinical and preclinical perspectives on NDDSs in oncology

It offers valuable insights into the potential impact of these compounds on cancer treatment. In preclinical studies, NDDSs have shown promising results in enhancing the therapeutic efficacy of anticancer agents, reducing systemic toxicity, and overcoming drug resistance. These studies have typically involved the use of animal models or in vitro models to evaluate the safety, pharmacokinetics, and effectiveness of NDDSs. Preclinical studies have also focused on understanding the mechanisms of drug release, targeting, and tumor penetration of NDDSs, which are crucial for optimizing their clinical translation (Singh et al., 2012; Villaverde and Baeza, 2019).

In clinical studies, NDDSs have been evaluated in cancer patients, focusing on safety, tolerability, and efficacy (Siddiqui et al., 2012). Clinical trials with promising results have explored the use of NDDSs in various cancer types, including solid tumors and hematologic malignancies (Saha et al., 2010). For instance, liposomal versions of chemotherapy drugs, such as doxorubicin and paclitaxel, have demonstrated better pharmacokinetics and fewer side effects compared to traditional formulations. Moreover, targeted drug delivery systems, such as ADCs and immunoconjugates, have shown improved tumor specificity and therapeutic outcomes (Cho et al., 2008).

Overall, both clinical and preclinical studies have highlighted the potential of NDDSs in improving cancer treatment. However, several challenges remain, including optimizing drug loading and release kinetics, achieving sufficient tumor penetration, and addressing heterogeneity within tumors. Future research endeavors should concentrate on resolving these issues to further optimize NDDSs for clinical use in oncology (Vasir et al., 2005; Shaffer, 2005).

Challenges and future directions in NDDSs for cancer management

NDDSs offer significant potential for enhancing cancer management, but several challenges must be addressed for their full realization (Ho et al., 2017). One critical issue is precisely targeting cancer cells while minimizing off-target effects. While NDDSs have progressed in this area through ligand-targeted delivery and stimuli-responsive systems, further refinement is necessary to ensure accurate and sustained drug delivery exclusively to cancer cells. Another challenge is effectively encapsulating and regulating the release of therapeutic agents. Even though NDDSs have proven capable of encapsulating various medications, obtaining uniform drug release profiles is still a concern, particularly for medicines with variable solubilities and stabilities (Mitra et al., 2015).

Moreover, ensuring the safety and biocompatibility of NDDSs is paramount, as these systems interact with the complex biological environment of the body. Developing and verifying strategies to reduce toxicity, immunogenicity, and other adverse effects of NDDSs is imperative (Pagar and Khandbahale, 2019). To maximize NDDS production for clinical use, scaling up is also necessary. Further standardizing and validating the production process is necessary to guarantee reproducibility, stability, and quality control. The manufacturing process must ensure reproducibility, stability, and quality control, which requires further standardization and validation (Li et al., 2023).

Looking to the future, several directions promise to advance NDDSs in cancer management. The integration of NDDSs with immunotherapy, gene therapy, and other emerging treatment modalities could lead to synergistic effects and improved patient outcomes. Furthermore, individualized NDDSs that are personalized to each patient's unique traits and tumor profile may improve treatment outcomes and reduce side effects. Moreover, real-time monitoring of medication administration and tumor response may be possible using biomarkers and sophisticated imaging techniques, which would help with therapy selection and enhance patient outcomes. In general, realizing the full potential of NDDSs in revolutionizing cancer management would require addressing present issues and taking advantage of upcoming opportunities (Yap et al., 2021).

Integrating the NDDSs with immunotherapy and precision medicine

Combining immunotherapy and precision medicine with NDDSs is a potent cancer treatment strategy. Using the body's immune system to target and eliminate cancer cells specifically, immunotherapy is potentially a more effective and long-lasting treatment option than conventional chemotherapy. In contrast, precision medicine focuses on personalizing cancer treatment to a patient's genetic profile, tumor features, and other specific parameters to maximize therapeutic success and minimize side effects. Further improving the results of cancer treatment is possible by combining NDDSs with immunotherapy and precision medicine. By improving the transport of immunotherapeutic agents—such as monoclonal antibodies or checkpoint inhibitors—to tumor sites, NDDSs can lessen systemic toxicity and increase efficacy. Moreover, NDDSs can be tailored to deliver personalized treatments, such as gene or RNA-based therapies, directly to cancer cells, exploiting their unique molecular characteristics. This approach offers the potential to overcome drug resistance mechanisms, optimize therapeutic outcomes, and pave the way for more personalized and effective cancer treatments.

Conclusion

Developing NDDSs represents a noteworthy advancement in cancer treatment, providing a versatile and efficient way to administer therapeutic drugs with increased accuracy and reduced complications. These systems, including liposomes, polymeric NPs, and micelles, have demonstrated promising results in preclinical and clinical settings, showing their potential to revolutionize cancer treatment. By encapsulating chemotherapeutic agents, immunotherapies, or gene therapies, NDDSs can improve drug stability, bioavailability, and tumor targeting, thereby enhancing therapeutic efficacy while minimizing systemic toxicity. Integrating NDDSs with immunotherapy and precision medicine holds further promise, offering personalized and targeted treatments tailored to individual patient needs and tumor characteristics. Subsequent investigations should address these issues and improve NDDSs for clinical cancer treatment. Nevertheless, NDDSs have the potential to revolutionize cancer care by creating new avenues for better patient outcomes and quality of life.

References

Anitha, K., Chenchula, S., Surendran, V., Shvetank, B., Ravula, P., Milan, R., Chikatipalli, R., Padmavathi, R., 2024. Advancing cancer theranostics through biomimetics: a comprehensive review. Heliyon. 2405-8440. 10 (6), e27692. doi:10.1016/j.heliyon.2024.e27692.

Begines, B., Ortiz, T., Pérez-Aranda, M., Martínez, G., Merinero, M., Argüelles-Arias, F., Alcudia, A., 2020. Polymeric nanoparticles for drug delivery: recent developments and future prospects. Nanomaterials. 2079-4991. 10 (7), 1403. doi:10.3390/nano10071403.

Brigger, I., Dubernet, C., Couvreur, P., 2002. Nanoparticles in cancer therapy and diagnosis. Adv. Drug Deliv. Rev. 0169-409X. 54 (5), 631–651. doi:10.1016/s0169-409x(02)00044-3.

Buwalda, S.J., Vermonden, T., Hennink, W.E., 2017. Hydrogels for therapeutic delivery: current developments and future directions. Biomacromolecules. 1525-7797. 18 (2), 316–330. doi:10.1021/acs.biomac.6b01604.

Chaturvedi, V.K., Sharma, B., Tripathi, A.D., Yadav, D.P., Singh, K.R.B., Singh, J., Singh, R.P., 2023. Biosynthesized nanoparticles: a novel approach for cancer therapeutics. Front. Med. Technol. 2673-3129. 5. Available from: https://www.frontiersin.org/journals/medical-technology. doi:10.3389/fmedt.2023.1236107.

Cho, K., Wang, X., Nie, S., Chen, Z., Shin, D.M., 2008. Therapeutic nanoparticles for drug delivery in cancer. Clin. Cancer Res. 1078-0432. 14 (5), 1310–1316. Available from: http://clincancerres.aacrjournals.org/cgi/reprint/14/5/1310. doi:10.1158/1078-0432.CCR-07-1441.

Di Stefano, A., 2023. Nanotechnology in targeted drug delivery. IJMS 24 (9), 8194. doi:10.3390/ijms24098194 .

Edis, Z., Wang, J., Waqas, M.K., Ijaz, M., Ijaz, M., 2021. Nanocarriers-mediated drug delivery systems for anticancer agents: an overview and perspectives. Int. J. Nanomed. 1178-2013. 16, 1313–1330. doi:10.2147/ijn.s289443.

Emerich, D.F., 2005. Nanomedicine - prospective therapeutic and diagnostic applications. Expert Opin. Biol. Ther. 1471-2598. 5 (1), 1–5. doi:10.1517/14712598.5.1.1.

Fornaguera, C., García-Celma, M.J., 2017. Personalized nanomedicine, a revolution at the nanoscale. J. Pers. Med. 7 (4), 12. doi:10.3390/jpm7040012.

Gagliardi, A., Giuliano, E., Venkateswararao, E., Fresta, M., Bulotta1, S., Awasthi, V., Cosco, D., 2021. Biodegradable polymeric nanoparticles for drug delivery to solid tumors. Front. Pharmacol. 12, 601626. doi:10.3389/fphar.2021.601626.

Garg, N.K., Tandel, N., Jadon, R.S., Tyagi, R.K., Katare, O.P., 2018. Lipid–polymer hybrid nanocarrier-mediated cancer therapeutics: current status and future directions. Drug Discov. Today 1878-5832. 23 (9), 1610–1621. Available from: https://www.elsevier.com/locate/drugdiscov. doi:10.1016/j.drudis.2018.05.033.

George, E., 2023. Advancements in targeted drug delivery systems for enhanced therapeutic outcomes. J. Pharm. Toxicol. 6 (3), 61–63.

Gu, Z., Da Silva, C.G., Van der Maaden, K., Ossendorp, F., Cruz, L.J., 2020. Liposome-based drug delivery systems in cancer immunotherapy. Pharmaceutics. 1999-4923. 12 (11), 1054. doi:10.3390/pharmaceutics12111054.

Hamid, R., Manzoor, I., 2020. Nanomedicines, Nano Based Drug Delivery Systems Challenges and Opportunities. IntechOpen, London, UK.

Hani, U., Rahamathulla, M., Osmani, R.A., Kumar, H.Y., Urolagin, D., Ansari, M.Y., Pandey, K., Devi, K., Yasmin, S., 2020. Recent advances in novel drug delivery systems and approaches for management of breast cancer: a comprehensive review. J. Drug Deliv. Sci. Technol. 1773-2247. 56, 101505. doi:10.1016/j.jddst.2020.101505.

Ho, B.N., Pfeffer, C.M., Singh, A.T.K., 2017. Update on nanotechnology-based drug delivery systems in cancer treatment. Anticancer Res. 1791-7530. 37 (11), 5975–5981. Available from: http://ar.iiarjournals.org/content/37/11/5975.full.pdf. doi:10.21873/anticanres.12044.

Ho, T.C., Chang, C.C., Chan, H.P., Chung, T.W., Shu, C.W., Chuang, K.P., Duh, T.H., Yang, M.H., Tyan, Y.C., 2022. Hydrogels: properties and applications in biomedicine. Molecules 1420-3049. 27 (9). Available from: https://www.mdpi.com/1420-3049/27/9/2902/pdf?version=1651484823. doi:10.3390/molecules27092902.

Hong, L., Li, W., Li, Y., Yin, S., 2023. Nanoparticle-based drug delivery systems targeting cancer cell surfaces. RSC Adv. 2046-2069. 13 (31), 21365–21382. doi:10.1039/d3ra02969g.

Hu, C.-M.J., Aryal, S., Zhang, L., 2010. Nanoparticle-assisted combination therapies for effective cancer treatment. Ther. Deliv. 2041-5990. 1 (2), 323–334. doi:10.4155/tde.10.13.

Khizar, S., Alrushaid, N., Khan, Zine, N., Jaffrezic-Renault, N., Errachid, A., Elaissari, A., 2023. Nanocarriers based novel and effective drug delivery system. Int. J. Pharm. 632, 122570. doi:10.1016/j.ijpharm.2022.122570.

Kumar, R., Saha, P., Sarkar, S., Rawat, N., Prakash, A., 2021. A review on novel drug delivery system. IJRAR 8, 183–199.

Kuna, K., Baddam, S.R., Kalagara, S., Akkiraju, P.C., Rahul, S., Tade, Enaganti, S., 2024. Emerging natural polymer-based architectured nanotherapeutics for the treatment of cancer. Int. J. Biol. Macromol. 0141-8130. 262 (Pt 1), 129434. doi:10.1016/j.ijbiomac.2024.129434.

Kydd, J., Jadia, R., Velpurisiva, P., Gad, A., Paliwal, S., Rai, P., 2017. Targeting strategies for the combination treatment of cancer using drug delivery systems. Pharmaceutics. 1999-4923. 9 (4), 46. doi:10.3390/pharmaceutics9040046.

Li, J., Wang, Q., Xia, G., Adilijiang, N., Li, Y., Hou, Z., Fan, Z., Li, J., 2023. Recent advances in targeted drug delivery strategy for enhancing oncotherapy. Pharmaceutics 15 (9), 2233. https://www.mdpi.com/1999-4923/15/9/2233.

Liu, Y., Castro Bravo, K.M., Liu, J., 2021. Targeted liposomal drug delivery: a nanoscience and biophysical perspective. Nanoscale Horiz. 2055-6756. 6 (2), 78–94. doi:10.1039/d0nh00605j.

Madaan, K., Kumar, S., Poonia, N., Lather, V., Pandita, D., 2014. Dendrimers in drug delivery and targeting: drug-dendrimer interactions and toxicity issues. J. Pharm. Bioallied Sci. 0975-7406. 6 (3), 139–150. Available from: http://www.jpbsonline.org. doi:10.4103/0975-7406.130965.

Malhi, M., Sharma, S.K., 2022. A study on biopharmaceutical considerations in controlled drug delivery. IJIEMR 11, 1601–1608.

Marwadi, M., Borade, P., Kandalkar, S., 2022. A review on novel drug delivery system. Int. J. Adv. Eng. Manag. 4, 1403–1411.

Mintzer, M.A., Grinstaff, M.W., 2011. Biomedical applications of dendrimers: a tutorial. Chem. Soc. Rev. 0306-0012. 40 (1), 173–190. doi:10.1039/b901839p.

Mitra, A.K., Agrahari, V., Mandal, A., Cholkar, K., Natarajan, C., Shah, S., Joseph, M., Trinh, H.M., Vaishya, R., Yang, X., Hao, Y., Khurana, V., Pal, D., 2015. Novel delivery approaches for cancer therapeutics. J. Control. Release 1873-4995. 219, 248–268. Available from: https://www.elsevier.com/locate/jconrel. doi:10.1016/j.jconrel.2015.09.067.

Montané, X., Bajek, A., Roszkowski, K., Montornés, J.M., Giamberini, M., Roszkowski, S., Kowalczyk, O., Garcia-Valls, R., Tylkowski, B., 2020. Encapsulation for cancer therapy. Molecules. 1420-3049. 25 (7), 1605. doi:10.3390/molecules25071605.

Nie, S., Xing, Y., Kim, G.J., Simons, J.W., 2007. Nanotechnology applications in cancer. Annu. Rev. Biomed. Eng. 1523-9829. 9, 257–288. doi:10.1146/annurev.bioeng.9.060906.152025.

Nilesh, K., Umama, Y., khan, M.G., Kushwah, M., KUMAR, A.Y.U.S.H., 2019. The fundamental of novel drug delivery system; methodology, role of nanotechnology; nanoparticles in pharmaceutical research. SSRN Electron. J. 1556-5068. doi:10.2139/ssrn.3557341.

Pagar, K.R., Khandbahale, S.V., 2019. A review on novel drug delivery system: a recent trend. Asian J. Pharm. Technol. 2231-5705. 9 (2), 135. doi:10.5958/2231-5713.2019.00023.0.

Puccetti, M., Pariano, M., Schoubben, A., Giovagnoli, S., Ricci, M., 2024. Biologics, theranostics, and personalised medicine in drug delivery systems. Pharmacol. Res. 201, 107086, doi:10.1016/j.phrs.2024.107086.

Rahim, M.A., Jan, N., Khan, S., Shah, H., Madni, A., Khan, A., Jabar, A., Khan, S., Elhissi, A., Hussain, Z., Aziz, H.C., Sohail, M., Khan, M., Thu, H.E., 2021. Recent advancements in stimuli responsive drug delivery platforms for active and passive cancer targeting. Cancers 2072-6694. 13 (4), 1–52. Available from: https://www.mdpi.com/2072-6694/13/4/670/pdf. doi:10.3390/cancers13040670.

Raj, S., Khurana, S., Choudhari, R., Kesari, K.K., Kamal, M.A., Garg, N., Ruokolainen, J., Das, B.C., Kumar, D., 2021. Specific targeting cancer cells with nanoparticles and drug delivery in cancer therapy. Semin. Cancer Biol. 1044-579X. 69, 166–177. doi:10.1016/j.semcancer.2019.11.002.

Saha, R.N., Vasanthakumar, S., Bende, G., Snehalatha, M., 2010. Nanoparticulate drug delivery systems for cancer chemotherapy. Mol. Membr. Biol. 0968-7688. 27 (7), 215–231. doi:10.3109/09687688.2010.510804.

Salahshoori, I., Golriz, M., Nobre, M.A.L., Mahdavi, S., Eshaghi Malekshah, R., Javdani-Mallak, A., Namayandeh Jorabchi, M., Ali Khonakdar, H., Wang, Q., Mohammadi, A.H., Masoomeh Sadat Mirnezami, S., Kargaran, F., 2024. Simulation-based approaches for drug delivery systems: navigating advancements, opportunities, and challenges. J. Mol. Liq. 0167-7322. 395, 123888. doi:10.1016/j.molliq.2023.123888.

Shaffer, C., 2005. Nanomedicine transforms drug delivery. Drug Discov. Today. 1359-6446. 10 (23-24), 1581–1582. doi:10.1016/s1359-6446(05)03654-8.

Sharma, A., Sharma, U.S., 1997. Liposomes in drug delivery: progress and limitations. Int. J. Pharm. 0378-5173. 154 (2), 123–140. doi:10.1016/S0378-5173(97)00135-X.

Siddiqui, I.A., Adhami, V.M., Chamcheu, C.J., Mukhtar, H., 2012. Impact of nanotechnology in cancer: emphasis on nanochemoprevention. Int. J. Nanomed. 1178-2013. 7, 591–605. Available from: https://www.dovepress.com/getfile.php?fileID=11965.

Singh, S., Sharma, A., Robertson, G.P., 2012. Realizing the clinical potential of cancer nanotechnology by minimizing toxicologic and targeted delivery concerns. Cancer Res. 1538-7445. 72 (22), 5663–5668. Available from: http://cancerres.aacrjournals.org/content/72/22/5663.full.pdf+html. doi:10.1158/0008-5472.CAN-12-1527.

Snehal, A., Gavhane, A., Gade, T., 2021. The novel drug delivery system. IJCRT 9, 373–385.

Sun, H., Zhang, Y., Wang, G., Yang, W., Xu, Y., 2023. mRNA-based therapeutics in cancer treatment. Pharmaceutics. 1999-4923. 15 (2), 622. doi:10.3390/pharmaceutics15020622.

Svenson, S., 2009. Dendrimers as versatile platform in drug delivery applications. Eur. J. Pharm. Biopharm. 71, 445–462.

Vaibhav, A., Borase, Kapadnis, R.S., Chaudhari, A.Y., Bachhav, S.R., 2020. A review on novel drug delivery system. JETIR 7, 19–27.

Vasir, J., Reddy, M., Labhasetwar, V., 2005. Nanosystems in drug targeting: opportunities and challenges. Curr. Nanosci. 1573-4137. 1 (1), 47–64. doi:10.2174/1573413052953110.

Verma, M., Deep, A., Nandal, R., Shinmar, P., Kaushik, D., 2017. Novel drug delivery system for cancer management: a review. Curr. Cancer Ther. Rev. 1573-3947. 12 (4), 253–272. doi:10.2174/1573394713666170406101900.

Villaverde, G., Baeza, A., 2019. Targeting strategies for improving the efficacy of nanomedicine in oncology. Beilstein J. Nanotechnol. 2190-4286. 10 (1), 168–181. doi:10.3762/bjnano.10.16.

Xin, Y., Yin, M., Zhao, L., Meng, F., Luo, L., 2017. Recent progress on nanoparticle-based drug delivery systems for cancer therapy. Cancer Biol. Med. 2095-3941. 14 (3), 228. doi:10.20892/j.issn.2095-3941.2017.0052.

Yap, T.A., Parkes, E.E., Peng, W., Moyers, J.T., Curran, M.A., Tawbi, H.A., 2021. Development of immunotherapy combination strategies in cancer. Cancer Discov. 2159-8274. 11 (6), 1368–1397. doi:10.1158/2159-8290.cd-20-1209.

Yezdani, Khan, G., Kushwah, N., Verma, A., Khan, F., 2018. Application of nanotechnology in diagnosis and treatment of various diseases and its future advances in medicine. World J. Pharm. Pharm. Sci. 7, 1611–1633.

Zielińska, A., Carreiró, F., Oliveira, A.M., Neves, A., Pires, B., Nagasamy Venkatesh, D., Durazzo, A., Lucarini, M., Eder, P., Silva, A.M., Santini, A., Souto, E.B., 2020. Polymeric nanoparticles: production, characterization, toxicology and ecotoxicology. Molecules. 1420-3049. 25 (16), 3731. doi:10.3390/molecules25163731.

CHAPTER

7

Palliative care in oncology: Symptom management and supportive therapies

Gautam Sarma[a], Jyotiman Nath[b], and Partha Pratim Medhi[a]

[a]Department of Radiation Oncology, AIIMS, Guwahati, Assam, India
[b]Department of Radiation Oncology, Dr. B. Borooah Cancer Institute, Guwahati, Assam, India

Introduction

Palliative care focuses on addressing the physical, psychological, social, and spiritual needs of individuals facing life-threatening illnesses. It enhances patients' overall well-being by providing comprehensive, personalized support to meet their unique challenges. Moreover, it plays a crucial role in uplifting and supporting caregivers, significantly improving their quality of life alongside that of the patients. The World Health Organization (WHO) estimates that each year, 56.8 million people, including 25.7 million in their final year of life, require palliative care. Yet, alarmingly, only 14% of those in need have access to these essential services, underscoring a profound global shortfall in palliative care availability (World Health Organization, 2025).

Palliative care in oncology is an essential component of comprehensive cancer treatment, focusing on improving the quality of life for patients and their families by addressing the multidimensional challenges associated with the disease. Cancer patients experience a broad spectrum of physical, emotional, and psychological challenges throughout the disease trajectory. Palliative care prioritizes symptom relief and holistic support, often provided alongside curative or life-prolonging treatments. Integrating palliative care into oncology has demonstrated improved outcomes, including enhanced quality of life, better symptom control, and even prolonged survival in some patient populations. For instance, a landmark study published in the *New England Journal of Medicine* in 2010 by Temel et al. (2010) showed that early integration of palliative care in patients with metastatic nonsmall-cell lung cancer resulted in significant improvements in both quality of life and mood, along with a median survival benefit of approximately 2 months compared to standard oncological care alone.

Pharmacological Approaches in Cancer Treatment
https://doi.org/10.1016/B978-0-443-38307-6.00007-7

This chapter provides an overview of the key aspects of palliative care in oncology, including pharmacological interventions, supportive therapies, and strategies for managing end-of-life (EOL) care. It highlights the importance of interdisciplinary collaboration and patient-centered approaches to address the unique needs of individuals with cancer.

Pain management

Cancer pain is one of the most distressing symptoms experienced by patients, significantly impacting their quality of life. A recent meta-analysis of 444 studies from 2014 to 2021 reported the overall prevalence of pain in cancer patients as 44%, with moderate to severe pain experienced by 30.6% of the patients (Snijders et al., 2023). Cancer pain can arise from tumor progression, treatment-related side effects, or a combination of both. Pain in cancer patients is classified as nociceptive, stemming from tissue damage, or neuropathic, resulting from nerve injury or dysfunction. Effective pain management, a cornerstone of palliative care, is crucial to alleviating suffering and enhancing functionality.

Pharmacological management of cancer pain

The WHO's three-step analgesic ladder provides a structured approach to managing cancer pain. The WHO's three-step analgesic ladder remains the cornerstone of cancer pain management, offering a structured approach to the management of cancer-related pain (WHO, 2018).

1. **Nonopioids**: Step I of the WHO ladder involves analgesic drug therapy for patients with mild to moderate pain. Such patients should be treated with nonopioid analgesics, particularly nonsteroidal antiinflammatory drugs, which may or may not be combined with an adjuvant drug, depending on the specific pain pathophysiology. These medications include acetaminophen, nonspecific cyclooxygenase inhibitors (including ibuprofen, diclofenac, and aspirin), and cyclooxygenase-2 inhibitors (Paget, 1989).
2. **Opioids**: Step II of the WHO ladder is indicated for patients with moderate pain who do not experience pain relief with Step I medications. These patients are candidates for a combination of a nonopioid and a low dose of opioid analgesics, such as codeine, oxycodone, or morphine. For moderate to severe pain, opioids like morphine, fentanyl, and oxycodone are pivotal. Dosage titration and regular monitoring help balance effective pain relief with minimal adverse effects. Step III is reserved for patients who report either severe pain (often gauged as greater than 7 on a 0–10 scale) or moderate pain that is inadequately managed after the appropriate administration of drugs at the second step of the WHO ladder. For these patients, nonopioids are often used in combination with more potent doses of opioids to mitigate the opioid effect, and adjuvants are administered depending on the pain pathophysiology or the need to control other concurrent symptoms in the individual patient.
3. **Adjuvant therapies**: These include antidepressants (e.g., amitriptyline, duloxetine) for neuropathic pain, anticonvulsants (e.g., gabapentin, pregabalin) for nerve-related pain,

corticosteroids for inflammation-induced pain and tumor-related edema. Sodium channel blockers, N-methyl-D-aspartate receptor antagonists, and cannabinoids may be used in the treatment of neuropathic pain. Interventional strategies, such as nerve blocks and spinal drug delivery systems, may be employed for refractory cases (WHO, 2018).

Nonpharmacological approaches

Nonpharmacological strategies complement pharmacological treatments, offering holistic care in cancer pain management.

- **Radiation therapy:** It has been reported to be effective in palliating pain due to metastatic bone disease. Overall response rates of 70%–80% are reported, with approximately one-third of patients achieving complete pain relief. Notably, 40% of responders experience pain relief within 10 days (Rich et al., 2018). The American Society for Radiation Oncology recommends using single-fraction external beam radiotherapy at a dose of 8 Gy for palliative pain management of bone metastases. Fractionated schedules often result in more significant patient inconvenience and higher costs than a single-fraction approach (Lutz et al., 2017). Updated robust data analyses confirm that pain relief is comparable across different regimens, including a single 8 Gy fraction, 20 Gy in 5 fractions, 24 Gy in 6 fractions, and 30 Gy in 10 fractions for patients with previously untreated painful bone metastases (Lutz et al., 2011).
- **Psychological interventions**: Several psychosocial factors have been implicated as barriers to successful pain palliation. Among these, emotional distress, particularly anxiety, depression, and beliefs about pain, has consistently emerged as a predictor of patient's pain levels. Supportive psychotherapy for cancer patients is performed in several formats, including individual, family, couples, and group settings, which provide emotional support and assist patients in adapting to the crisis. Cognitive-behavioral therapy is used to modify physiological pain reactions and behaviors. This is achieved by helping patients learn relaxation techniques, which further help improve self-control, divert attention, and break the pain–anxiety–tension cycle (Thomas and Weiss, 2010; Fallon et al., 2018).
- **Advanced interventional techniques:** Recent adaptations of the WHO ladder include a fourth step, incorporating advanced interventions such as intrathecal drug delivery, neuromodulation, and targeted radiotherapy for refractory cases. These techniques are particularly valuable for patients with complex or treatment-resistant pain. Innovations in cancer pain taxonomy, such as the inclusion of chronic cancer pain in ICD-11, underscore the evolving understanding of cancer-related pain. This classification aids in public health planning and ensures greater global visibility for cancer pain management needs. A multidisciplinary, integrative approach that aligns with patient-centered care principles remains essential for advancing cancer pain management and improving the quality of life for patients and their families (Mestdagh et al., 2023).

Management of other symptoms

Cancer often leads to a constellation of symptoms that significantly impact patients' well-being. Effective management involves identifying underlying causes and implementing targeted interventions.

Nausea and vomiting

Nausea and vomiting are distressing symptoms commonly experienced by cancer patients, particularly as side effects of chemotherapy, radiotherapy, or advanced disease. If left untreated, these symptoms can significantly impair nutritional intake, hydration status, and overall quality of life. Effective management involves identifying the underlying etiology and implementing a combination of pharmacological and nonpharmacological strategies.

Etiology

The causes of nausea and vomiting in oncology patients are multifactorial, with chemotherapy being the most common trigger. Other contributing factors include disease-related complications, metabolic imbalances such as hypercalcemia or renal failure, and gastrointestinal (GI) obstruction due to tumor burden. Increased intracranial pressure from brain metastases and vestibular dysfunction caused by central nervous system involvement can also lead to these symptoms. In addition, radiation-induced nausea and vomiting is a significant concern in patients undergoing radiation therapy, particularly when targeting the abdomen, brain, or GI tract (Keeley, 2009; Davis and Hallerberg, 2010; Delgado-Guay et al., 2009; Gupta et al., 2013).

Chemotherapy-induced nausea and vomiting

Chemotherapy-induced nausea and vomiting (CINV) is one of the most distressing and prevalent side effects of cancer treatment, affecting up to 40% of patients. It is not only a significant physical burden but also one of the most feared complications among those undergoing chemotherapy. Despite advancements in antiemetic therapies, CINV remains a major challenge, often impairing patients' quality of life, nutritional status, and overall treatment adherence (Dranitsaris et al., 2017; Lorusso et al., 2017; de Boer-dennert et al., 1997).

CINV is classified into distinct types, each requiring specific management strategies. Acute CINV occurs within the first 24 h after chemotherapy administration and is primarily mediated by serotonin release. Delayed CINV develops after 24 h and can persist for several days, often driven by substance P and NK1 receptor activation. Anticipatory CINV is a conditioned response triggered by previous negative experiences with chemotherapy, leading to nausea and vomiting even before treatment begins. Breakthrough CINV refers to symptoms that occur despite the use of prophylactic antiemetic therapy, requiring additional intervention. Lastly, Refractory CINV describes persistent nausea and vomiting that remain uncontrolled despite optimized antiemetic regimens across multiple chemotherapy cycles, presenting a significant challenge in symptom management (Aapro, 2018; Janelsins et al., 2013; Navari et al., 2013).

Pathophysiology of CINV

CINV is triggered by complex interactions between the central and peripheral pathways of the nervous system, primarily involving the vomiting center (VC) in the medulla oblongata.

Peripheral Pathway: Chemotherapy drugs damage enterochromaffin cells in the GI tract, leading to the release of serotonin (5-HT3). This activates vagal afferent nerves, which send signals to the dorsal vagal complex, ultimately stimulating the vomiting reflex. This pathway is mainly responsible for acute CINV (occurring within 24 h of chemotherapy) (Hesketh, 2008; Shankar et al., 2015).

Central pathway: The chemoreceptor trigger zone (CTZ), located in the area postrema of the brainstem, detects circulating toxins, including chemotherapy agents, and relays signals to the VC. Neurotransmitters such as substance P (acting on NK1 receptors), dopamine (D2 receptors), serotonin (5-HT3 receptors), and histamine contribute to the emetic response. This pathway is primarily associated with delayed CINV that occurs after 24 h (Herrstedt et al., 2005; Andrews and Sanger, 2002).

Management of CINV

International antiemetic guidelines provide specific recommendations on the selection of antiemetic agents for CINV prevention, tailored to the emetogenic potential of the chemotherapy regimen. A multimodal approach is recommended, combining pharmacological and nonpharmacological interventions to optimize control (National Comprehensive Cancer Network (NCCN), 2024a; Glare et al., 2011; Smith et al., 2012).

Pharmacological management

5-HT3 receptor antagonists

5-HT3 receptor antagonists work by blocking serotonin-mediated activation of the VC, thereby preventing CINV. Common examples include ondansetron, granisetron, and palonosetron, with palonosetron being particularly effective for delayed CINV due to its longer half-life. These agents are highly effective in managing acute CINV, occurring within the first 24 h postchemotherapy, but their efficacy in controlling delayed CINV is comparatively lower, often necessitating combination therapy with other antiemetics such as NK1 receptor antagonists or corticosteroids (Doggrell, 2017).

NK1 receptor antagonists

NK1 receptor antagonists prevent nausea and vomiting by blocking the binding of substance P to NK1 receptors in the brain. Common agents in this class include aprepitant, fosaprepitant, and rolapitant. These drugs are primarily used in combination with 5-HT3 receptor antagonists and corticosteroids to effectively manage delayed CINV, which occurs >24 h after chemotherapy administration (Adel, 2017; Rao and Faso, 2012; Rojas et al., 2014).

Corticosteroids

Corticosteroids, particularly dexamethasone, play a crucial role in enhancing the efficacy of antiemetic regimens. They are effective in managing both acute and delayed CINV, often used in combination with 5-HT3 receptor antagonists and NK1 receptor antagonists to provide optimal symptom control. Their antiinflammatory and antiemetic properties contribute to reducing CINV (Adel, 2017; Rao and Faso, 2012; Chu et al., 2014).

Dopamine antagonists

Dopamine antagonists, such as metoclopramide, prochlorperazine, and haloperidol, are commonly used as rescue therapy for breakthrough CINV: nausea and vomiting that occur despite prophylactic antiemetic treatment. By blocking dopamine receptors in the CTZ, these agents help manage persistent symptoms when standard regimens are insufficient (Roila et al., 2016; Tan et al., 2009; Navari, 2014).

Adjunctive therapies

Atypical antipsychotics, benzodiazepines, and cannabinoids play important roles in managing CINV, particularly in cases where standard antiemetic regimens are insufficient. Olanzapine, an atypical antipsychotic, is gaining prominence due to its broad receptor activity, making it highly effective for breakthrough and refractory CINV, as well as delayed nausea. Benzodiazepines, such as lorazepam and alprazolam, are primarily used for anticipatory CINV, helping to reduce anxiety and prevent nausea triggered by previous negative chemotherapy experiences. Cannabinoids, including dronabinol and nabilone, are reserved for refractory cases where conventional treatments fail; however, their use is limited due to central nervous system side effects such as dizziness, sedation, and cognitive impairment (Roila et al., 2016; Tan et al., 2009; Navari, 2014).

Nontreatment-related nausea and vomiting in advanced cancer

Nausea and vomiting in cancer patients are not always related to chemotherapy or radiation therapy. Several other physiological and behavioral factors contribute to these symptoms, necessitating targeted interventions. Physiological causes include brain metastases, which may require radiation therapy or dexamethasone, and GI conditions such as gastritis or gastroesophageal reflux disease, which are often managed with H2 blockers or proton pump inhibitors. Dehydration due to inadequate fluid intake or systemic illness can be corrected with intravenous hydration, while electrolyte disturbances, such as hyponatremia or hypercalcemia, may require intravenous saline, diuretics, or bisphosphonates. Gastroparesis, a condition of delayed gastric emptying, can be alleviated with metoclopramide, whereas bowel obstruction may necessitate treatment with octreotide or antiemetics. Other contributing factors include vestibular dysfunction, which responds to meclizine; excessive secretions in head and neck cancer patients, managed with anticholinergics; and malignant ascites, which can be addressed through paracentesis or spironolactone. In addition, appetite loss and anorexia may benefit from olanzapine, megestrol acetate, or cannabinoids (Hardy and Davis, 2021; Ang et al., 2010; Glare et al., 2004).

Behavioral factors also play a role in nausea and vomiting. Anticipatory nausea, driven by patient expectations, is closely linked to anxiety and psychological distress, which may be alleviated with benzodiazepines like lorazepam and alprazolam. Depression-related nausea may respond to antidepressant therapy, while psychogenic nausea often requires psychiatric interventions (Kamen et al., 2014; Tremblay et al., 2003).

A particularly challenging clinical problem is chronic nausea and vomiting in cancer patients unrelated to chemotherapy or radiation. These symptoms can persist weeks after treatment completion, and their exact cause is often unclear. The recommended interventions for nontreatment-related nausea and vomiting are based on established guidelines, including those from the National Comprehensive Cancer Network (NCCN), MASCC/ESMO, and American Society of Clinical Oncology (ASCO) (National Comprehensive Cancer Network (NCCN), 2024a; Glare et al., 2011; Smith et al., 2012).

Nausea and vomiting remain significant challenges in oncology, affecting patients undergoing chemotherapy, radiation, and advanced disease progression. While CINV can be managed through evidence-based antiemetic strategies targeting key neurotransmitter pathways, nontreatment-related causes also require a comprehensive, multidisciplinary approach. The integration of pharmacological agents, behavioral interventions, and supportive care is essential for optimizing symptom control and improving the quality of life for cancer patients. Continued research and adherence to international guidelines will further refine management strategies, ensuring effective and individualized care.

Fatigue

Fatigue is one of the most common and distressing side effects of cancer and its treatment, often present before treatment and worsening with radiation, chemotherapy, hormonal, or biological therapies. Its prevalence during treatment varies widely, ranging from 25% to 99%, depending on patient population, treatment type, and assessment method (Jacobsen et al., 1999; Phillips et al., 2013). The NCCN has defined CRF as "a distressing, persistent, subjective sense of physical, emotional, and/or cognitive tiredness or exhaustion related to cancer or cancer-related treatment that is not proportional to recent activity and interferes with usual functioning" (National Comprehensive Cancer Network (NCCN), 2024b). Unlike ordinary fatigue, CRF is not alleviated by rest and can severely impact daily functioning. Despite its high prevalence, CRF is underreported by patients and undertreated by clinicians, largely due to the lack of standardized diagnostic criteria, variability in international guidelines, and inconsistencies in assessment methods. Research suggests that inflammation plays a key role in its pathophysiology, with certain patients at higher risk based on pretreatment factors. Effective management requires a comprehensive, evidence-based approach, integrating exercise, psychosocial interventions, and pharmacological therapies. However, barriers in recognition and treatment implementation remain significant challenges. Recent randomized controlled trials have highlighted the importance of tailored interventions, emphasizing physical activity, mind-body therapies, and targeted pharmacologic strategies to mitigate CRF and improve patient outcomes (Bower, 2014).

Pharmacological interventions of cancer-related fatigue

Cancer-related fatigue (CRF) is a multifactorial symptom that significantly impacts the quality of life and functional status in cancer patients. Despite its high prevalence, no single pharmacological agent has been universally effective, and treatment is often tailored to the underlying cause and patient-specific needs. Numerous studies have explored various pharmacological agents for CRF, including psychostimulants, corticosteroids, nutritional supplements, and hormonal therapies. However, the evidence supporting any one of these treatments remains limited and weak, often due to high heterogeneity between studies and the underrepresentation of advanced cancer patients in clinical trials. The available pharmacological options primarily aim to enhance energy levels, address anemia, modulate inflammatory pathways, and alleviate psychological distress, but a multimodal approach remains essential for optimal symptom management (Andreas et al., 2023; Pearson et al., 2018).

Psychostimulants

Psychostimulants, such as methylphenidate, improve alertness and cognitive function in CRF, particularly in chemotherapy and advanced disease patients, though side effects like insomnia and anxiety may limit their use. Modafinil and armodafinil, used for chemotherapy- and opioid-related fatigue, have shown limited benefit over placebo in recent trials, highlighting the need for further research (Andreas et al., 2023; Pearson et al., 2018).

Erythropoiesis-stimulating agents

Anemia is a major contributor to CRF, particularly in patients undergoing chemotherapy or radiation therapy. Darbepoetin and epoetin alfa stimulate red blood cell production, effectively reducing fatigue in anemic patients. However, their use is limited due to risks of thromboembolism, hypertension, and potential tumor progression, requiring careful patient selection. Iron and vitamin supplementation may benefit patients with iron deficiency anemia, particularly when combined with erythropoiesis-stimulating agents (Andreas et al., 2023; Minton et al., 2014).

Corticosteroids

Corticosteroids provide short-term symptomatic relief in patients with advanced CRF by reducing inflammation and increasing appetite and energy levels. Dexamethasone and prednisone are frequently used in palliative settings, offering temporary improvement in fatigue, but long-term use is discouraged due to adverse effects, including muscle wasting, hyperglycemia, and immunosuppression (Bruera et al., 2013; Moraska et al., 2010).

Hormonal and metabolic agents

Melatonin is involved in circadian rhythm regulation; melatonin supplementation has shown mixed results in improving fatigue, particularly in patients with sleep disturbances. Testosterone replacement therapy may benefit male patients with low testosterone levels, as hypogonadism has been linked to fatigue in cancer patients. L-Carnitine is a mitochondrial metabolism modulator investigated for its role in CRF, but clinical evidence remains inconclusive (Andreas et al., 2023; Cruciani et al., 2012)

Nonpharmacological strategies

Nonpharmacological strategies play a crucial role in alleviating CRF, offering safe and effective approaches that complement pharmacological treatments. Among these, exercise interventions have demonstrated the strongest evidence, with aerobic exercise, resistance training, and yoga consistently reducing fatigue levels in cancer patients during and after treatment. Studies suggest that even low-intensity activities, like walking, can improve energy levels, functional capacity, and overall well-being (Puetz and Herring, 2012; Strasser et al., 2013).

Cognitive behavioral therapy (CBT) and psychoeducational interventions are also proven methods to address fatigue, particularly when associated with anxiety, depression, or maladaptive coping behaviors. CBT helps restructure negative thought patterns, while psychoeducational approaches empower patients with self-management strategies. In addition, mindfulness-based interventions, relaxation techniques, and sleep hygiene training can significantly benefit patients struggling with persistent fatigue, insomnia, or psychological distress (Moyer et al., 2009; Jacobsen et al., 2007; Kangas et al., 2008; Goedendorp et al., 2009; Yates et al., 2005; Armes et al., 2007).

Emerging evidence also supports complementary therapies, such as acupuncture, massage therapy, and bright light therapy, though findings remain mixed and require further validation. Ultimately, a multimodal approach, integrating physical activity, psychological support, and lifestyle modifications, is essential for effective fatigue management in cancer patients (Andreas et al., 2023).

Dyspnea

Dyspnea or breathlessness is defined as a "subjective experience of breathing discomfort that consists of qualitatively distinct sensations that vary in intensity. The experience results from interactions among multiple physiological, psychological, social, and environmental factors and may induce secondary physiological and behavioral responses" (Dyspnea, 1999). It is a very distressing symptom for patients under palliative care. The frequency and intensity of breathlessness generally increase over the course of disease progression and ultimately reach their peak at death. In cancer, the prevalence of dyspnea is 50%–70% and up to 90% in patients with lung cancer (Rogers et al., 2025). The presence of comorbidities like congestive cardiac failure (CCF) and chronic obstructive pulmonary disease (COPD) in cancer patients often exacerbates the incidence and severity of dyspnea.

The etiology of dyspnea is a respiratory supply-and-demand mismatch. The mismatch can be due to increased oxygen demand, decreased lung capacity, or both. Respiration is maintained by regular rhythmical activity from the respiratory center in the brainstem, with inputs from the peripheral stretch receptors (airways, lung parenchyma, intercostal muscles, and diaphragm) and chemoreceptors (aortic and carotid bodies, medulla). Although the pathophysiology of dyspnea is not clearly understood, it is thought that in malignant diseases affecting the lung, dyspnea occurs primarily due to abnormalities in the stimulation of the peripheral stretch receptors, often with normal blood gas levels (Rogers et al., 2025; Carrieri et al., 1984).

Management of dyspnea begins with a thorough history taking and physical examination of the patient. A careful history must be elicited regarding onset, intensity, duration, diurnal variations, precipitating and relieving factors. A history of COPD and CCF must be enquired about, and any ongoing medications should be evaluated. Another important aspect is the association of dyspnea with anxiety and panic, so specific history regarding this aspect should be elicited. The management of specific conditions leading to dyspnea in cancer patients is highlighted in the following:

1 **Malignant pleural effusion (MPE):** MPE is reported in 15% of cancer patients and is secondary to lung cancer in about 35% of cases (Rodriguez-Panadero et al., 1989; Detterbeck et al., 2001). MPE patients present with varying degrees of respiratory stress. The degree of dyspnea depends on the effusion volume and the relationship between the lung and pleura. The short-term goal of therapy in MPE patients is the relief of dyspnea, while the long-term goal is to prevent fluid recollection by obliterating the pleural space (Aydin et al., 2009).

Therapeutic thoracocentesis is the most commonly employed therapy to relieve immediate symptoms of MPE. A maximum of 1500 cc should be evacuated immediately to avoid reexpansion pulmonary edema (Aydin et al., 2009). The disadvantage of thoracocentesis is the recollection of fluid over time, and hence, it is advisable in patients who are terminally ill with poor performance status and for whom no further therapy is possible. Also, care must be taken to avoid pneumothorax while carrying out the procedure. Intercostal tube drainage and pleurodesis are the most effective therapies for MPE and have the least chance of morbidity (Light, 1995). Small-bore tubes (8–14 F) are more comfortable for patients. Pleurodesis by intrapleural instillation of a sclerosant agent into the hemithorax obliterates the small pleural blood vessels, resulting in mesothelial fibrosis and collapse of the space between the visceral and parietal pleura. Chemical pleurodesis is the most commonly used method in MPE treatment. It is advisable to delay pleurodesis until drainage per day is <150 cc, and the chest tube is withdrawn when daily drainage is <100 cc (Lee et al., 2003b). The frequently used intrapleural sclerosing agents for pleurodesis are talc (Mg3Si4O10(OH)2), tetracycline hydrochloride, and bleomycin. Some centers also employ an indwelling pleural catheter to reduce symptoms and hospital stays while treating recurrent MPE effectively (Putnam et al., 1999). In patients who are refractory to chemical pleurodesis, pleuroperitoneal shunting is an alternative palliative therapy that may be considered. It comprises a subcutaneous pump that transfers the pleural fluid into the peritoneal cavity. Pleurectomy is another effective but invasive method that offers relief in chronic refractory MPE. It is reserved only for selected patients with failure of chemical pleurodesis in whom the malignant tissues have formed a cortex over the pleural surfaces. The risk of perioperative mortality is higher in this procedure as compared to other modalities of therapy (Fry and Khandekar, 1993). The prognosis and life expectancy of the patient versus the potential complications of the treatment must be considered when choosing the therapy modality for relief of breathlessness due to malignant MPE.

1. **Pulmonary thromboembolism (PTE):** Thromboembolism is one of the leading causes of morbidity and mortality in cancer patients, with the highest incidences reported in gastroenterological, hematological, pulmonary, prostatic, and gynecological cancers (Lee et al., 2021). The gold standard for diagnosing PTE is a CT pulmonary angiography, but in palliative care, most patients may not be fit to undergo such an investigation. Anticoagulation with low-molecular-weight heparin has been proven to be the first line of therapy for PTE after the publication of the landmark CLOT study (Lee et al., 2003a). However, in the palliative care setting, the benefits of anticoagulation must be weighed carefully against the risks of bleeding, relapse, and the presence of other comorbidities (e.g., thrombocytopenia, cardiac issues, etc.) when deciding on the initiation, continuation, and duration of treatment. In terminally ill patients, symptom palliation should be prioritized over definitive therapy (Chin-Yee et al., 2019).
2. **Superior vena caval obstruction (SVCO):** SVCO in cancer patients requires a thorough examination of the extent and location of the obstruction and the tissue histology of the tumor to decide on treatment. For chemotherapy and/or radiotherapy-sensitive malignancies (e.g., lymphoma, lung cancer, breast cancer, etc.), specific treatment therapy may be initiated, considering the patient's life expectancy and general condition. However, patients who have failed these therapies and/or have recurrent disease posttreatment may be candidates for thrombolysis (if the thrombus is the etiology), endovascular stenting, or surgical bypass, which may provide rapid relief. Often, in the terminal stages of palliative care, patients are unlikely to tolerate these procedures and are managed with supportive care such as steroid therapy (dexamethasone 8–16 mg/day), oxygen therapy, pain medications, diuretics, and nonpharmacological interventions (Chow et al., 2024; Nicholson et al., 1997).
3. **Lymphangitis carcinomatosis:** It is malignant infiltration and inflammation of the pulmonary interstitial lymphatic vessels secondary to the metastatic spread of malignancy from a primary site. It is generally indicative of a terminal stage in cancer patients. Steroids and antibiotics are routinely prescribed, although there is no scientific evidence to support their efficacy. Opioids and benzodiazepines may relieve dyspnea arising from this condition (Kumar and Mantri, 2025).
4. **Complications of cancer treatment:** Cancer treatment therapy (chemotherapy, radiotherapy, immunotherapy) may result in adverse reactions, anemia, respiratory infections, etc., that may lead to dyspnea. Careful attention must be given to the treatment history and ongoing definitive and supportive therapy to rule out any predisposing medications. In the terminal stages, when treatment therapy complications outweigh the benefits of survival, such therapies should be judiciously withheld.

Pharmacological treatment of dyspnea in terminally ill patients

1. **Opioids:** They remain the mainstay of treatment for refractory dyspnea in palliative care patients (Jennings et al., 2002; Zemel, 2022). They reduce respiratory rate by reducing the central perception of breathlessness, as well as reducing anxiety and

sensitivity to hypercapnia, although the exact mechanism is not clearly understood (Zemel, 2022).

Among all opioids, morphine has the most robust evidence for palliating dyspnea in patients with advanced cancer. However, the roles of other opioids, like oxycodone and fentanyl, as morphine substitutes in palliating dyspnea are emerging with recent evidence (Yamamoto et al., 2018; Hui et al., 2014). The most preferred routes of opioids for the relief of dyspnea are oral, subcutaneous, and parenteral. The benefit of nebulized opioids is doubtful for the alleviation of dyspnea. Although the exact dose of opioids in malignancy-induced dyspnea requires further research, it has been observed that opioids, even when prescribed in lower quantities, provide a reduction in breathlessness for patients with advanced disease, although not specifically in malignancy. A dosage of 20.1 ± 8.1 mg/day, with a median of 18 mg and a range of 5–36 mg/day, provided adequate relief to patients with decreased dyspnea without causing any serious side effects (Hui et al., 2014).

1 **Benzodiazepines:** They act by activating the inhibitory GABA pathway in the central nervous system. They mainly function as sedatives, hypnotics, muscle relaxants, anxiolytics, and anticonvulsants. Their effectiveness as monotherapy is debatable, but they serve as potent adjunctive therapy to opioids in the management of refractory dyspnea (Navigante et al., 2006; Clemens and Klaschik, 2011). Midazolam (10 mg/day subcutaneously) and lorazepam (0.5 mg sublingually) are most commonly employed as they are short-acting. Diazepam should be avoided in the management of dyspnea.
2 **Corticosteroids:** These are another group of agents that have shown adjunctive benefits along with opioids in some randomized trials (Maeda and Hayakawa, 2017). However, meta-analyses have shown that their role is inconclusive (Haywood et al., 2019).
3 **Saline, nebulized:** Sometimes employed to loosen thick secretions in the airway, but there is no evidence supporting its use in refractory dyspnea patients.
4 **Loop diuretics, nebulized:** Frusemide 40 mg inhaled by nebulization is beneficial in reducing induced dyspnea in healthy individuals (Grogono et al., 2018). However, its role in malignancy-induced dyspnea in palliative care patients remains controversial.
5 **Oxygen therapy:** Oxygen therapy in patients with dyspnea decreases the oxygen drive emanating from the brain's respiratory center (Ryan et al., 2014). However, oxygen is not beneficial in relieving dyspnea in both normoxic and hypoxic dyspneic patients with advanced cancer, as compared to the efficacy of opioids, which decrease both dyspnea severity and respiratory rates (Clemens et al., 2009). The Oxygen Trial randomized dyspnea patients to either room air or oxygen administration via nasal cannula. They reported that patients felt a decrease in breathlessness, irrespective of whether oxygen or room air was given via nasal cannula (LeBlanc and Abernethy, 2014). The trial instead suggests the potential advantages of facial and nasal airflow for dyspnea relief.

Nonpharmacological treatment of dyspnea in terminally ill patients

Dyspnea has a psychological aspect associated with its incidence and severity, especially in terminally ill advanced cancer patients. Some of the nonpharmacological therapies that have been studied in research and are worth mentioning are listed in the following (Puspawati et al., 2017; Swan et al., 2019; Bauml et al., 2016; Minchom et al., 2016; Yates et al., 2013; Gallagher et al., 2018; Guerrero-Torrelles et al., 2017):

1 **Facial airflow**: Airflow localized to the patient's face, either by medical technology or simply by a hand-held fan, stimulates the nervous system's perception of respiration and may relieve dyspnea.
2 **Acupuncture and acupressure**: Safe, although with mild utility.
3 **Breathing exercises and breathing training**: Inspiratory muscle training, pursed-lip breathing, diaphragmatic breathing, awareness of positioning, mindfulness-yogic breathing, and pacing breathing are some of the techniques suggested in the literature.
4 **Music therapy**: unequivocally acceptable to both the patient and caregivers.
5 **Spirituality**: Interventions to improve understanding of the meaning of our lives have been shown to enhance terminally ill patients' quality of life, self-empowerment, and ability to cope with physical suffering.

Anxiety and depression

Cancer patients go through a turmoil of mental stress during their journey from diagnosis to cure or death. They are expected to be prone to anxiety and depression, especially for those in the terminal stages of the disease and receiving palliative care. The prevalence of anxiety is 29%–52%, and the prevalence of depression is 27% (Gontijo Garcia et al., 2023). Studies have shown that various underlying factors related to cancer and its treatment predispose patients to symptoms of anxiety and depression (Kolva et al., 2011). Some of these factors are given as follows:

1 Fear of dying: common knowledge that cancer is a terminal illness and will eventually lead to an early death.
2 Disease progression: physical and mental burden associated with the failure of therapy.
3 Treatment-related symptoms: pain, fatigue, swallowing difficulties, nausea, etc., can directly impact mental health.
4 Loss of functionality: inability to carry out daily chores and acts of self-care may hamper the self-confidence of patients.
5 Uncertain future: feeling of hopelessness and uncertainty about whether the treatment will be successful.

There are many symptoms related to anxiety and depression, and not all of them may be seen in palliative care patients. However, the common symptoms can include excessive worry, restlessness, difficulty concentrating, irritability, and autonomic symptoms like rapid heart rate or sweating (Anon, 2019). Various factors exacerbate the symptoms, e.g., older individuals, patients with poor disease and symptom control, lack of family support, lower levels of physical, emotional, and cognitive functionality, etc. (Gontijo Garcia et al., 2023). The most important aspect of providing adequate support to such patients is the correct assessment of the level of anxiety and depression in them using validated tools and questionnaires. The ASCO, along with the American College of Surgeons Commission on Cancer and the NCCN, has issued guidelines stating that every cancer patient should be screened for anxiety at the initial cancer diagnosis and periodically thereafter (Andersen et al., 2014). Repeat assessments should be made whenever there are changes in cancer or treatment status (e.g., posttreatment, recurrence, or progression), as well as at transitions of the intention of treatment (e.g., definitive to palliative to hospice care). A brief workflow of screening palliative care patients for anxiety and depression, using the above-mentioned guideline, is shared as follows:

1 Initial assessment using the brief **Patient Health Questionnaire for Depression and Anxiety (PHQ-4),** a four-question-based patient questionnaire with a scale of 0–3 indicating increasing severity of symptoms experienced.
2 ***If the PHQ-4 screens positive for anxiety (or symptoms suggesting underlying anxiety are present),*** the seven-question-based **Generalized Anxiety Disorder (GAD-7) scale** needs to be employed (Spitzer et al., 2006). The GAD-7 scale, used to diagnose anxiety disorders and monitor symptoms over the course of the disease, is based on the Diagnostic and Statistical Manual of Mental Disorders, Fifth Edition (DSM-5), issued by the American Psychiatric Association (American Psychiatric Association, 2013).
3 ***If the PHQ-4 screen is positive for depression,*** DSM-5 criteria for assessing major depression or dysthymia should be used. If major depression is present, the nine-item **Patient Health Questionnaire (PHQ-9)** should be used to monitor its severity over the course of the disease and treatment.

Management of anxiety and depression in palliative care patients

Symptoms of anxiety require treatment when a GAD-7 score is 10 or more, and symptoms have started to interfere with daily activities or quality of life. The best way is to develop a strong, supportive relationship with the patient through a personalized and meaningful treatment approach for the patient and the clinical team (Anon, 2019). Treatment approaches include psychotherapy, psychopharmacologic treatments, and complementary treatments (exercise, lifestyle modifications, etc.).

1 **Psychotherapy:** It offers patients and their families an opportunity to explore the issues resulting in and exacerbating their anxiety symptoms, issues which cannot be tackled by medications alone. Some of these issues faced by cancer patients, which are commonly addressed via psychotherapy, include fears of medical interventions, physical symptoms, social impacts, finances, family support, future disability, dependency, spiritual issues, existential suffering, grief and loss, EOL issues, and dying (Anon, 2019; Storey and Knight, 1997). It should always be utilized to treat mental health issues in cancer patients, along with pharmacotherapy. Psychotherapy can be delivered individually, within families, or among groups. Some of the proven techniques of psychotherapy to improve quality of life by symptom control in palliative care patients are
   a Brief individual approach (Lo et al., 2014)
   b Cognitive behavioral therapy (Akechi et al., 2008; Greer et al., 2012)

c Supportive-expressive group therapy (Kissane et al., 2007)
d Dignity therapy (Chochinov et al., 2011)
e Meaning-centered psychotherapy (Breitbart et al., 2012)

2 **Pharmacotherapy:** Benzodiazepines are generally the first step in treating symptoms of anxiety or depression in cancer patients with medications (Storey and Knight, 1997, Storey and Knight, 1997). However, extreme caution should be used, as benzodiazepines cause increased toxicities in medically ill patients. Short-acting formulations like lorazepam are preferred in these situations, with careful monitoring of side effects (sedation, delirium, falls, etc.). If symptoms do not improve or are severe at onset, antidepressants or alternative agents may be employed. Serotonin–norepinephrine reuptake inhibitors (SNRIs) like venlafaxine and duloxetine may be employed as alternative antidepressants. Patients with a history of chronic anxiety or combined anxiety and depression may be initiated with a selective serotonergic reuptake inhibitor (SSRI) (Storey and Knight, 1997). Paroxetine is often the SSRI of choice in palliative care patients. For patients with severe anxiety, often combination therapy with both a benzodiazepine and an SSRI may have to be initiated, with slow withdrawal of the benzodiazepine once the therapeutic effect of the SSRI is reached.
3 **Complementary treatments:** Physical exercise has been shown to decrease worry and anxiety while improving overall functioning, even in palliative care patients. Even bedridden patients should be encouraged to use passive ranges of motion (Anon, 2019). Potential coping mechanisms, such as abstaining from caffeine, alcohol, and other intoxicating substances, should be practiced by patients undergoing palliative care. Maintenance of a sleep hygiene protocol may also help alleviate anxiety.
4 **Investigational agents:** 3,4-Methylenedioxymethamphetamine and psilocybin have been recently FDA-approved for the treatment of posttraumatic stress disorder and refractory depression (Kolbrich et al., 2008; Bogenschutz et al., 2022). Some small trials are encouraging their use as anxiolytics and antidepressants in terminally ill cancer patients too, but they require further detailed research.

It must also be borne in mind by the palliative care team that there are certain situations when mental health conditions might impact the delivery of primary treatment for the cancer patient. A referral to a mental health expert or psychiatrist is warranted in such clinical situations. Other scenarios where a psychiatry referral should be sought include significant worsening of symptoms despite medications, persistent anxiety lasting more than a week, and anxiety that interferes with the patient's ability to function.

Cancer cachexia

Cachexia is considered a metabolic syndrome caused by a chronic underlying disease. According to the Society of Cachexia and Wasting Disorders, it is defined as 5% weight loss in the previous 6 months with at least three of five clinical symptoms: fatigue, anorexia, reduced muscle strength, reduced fat-free mass, and/or systemic signs of inflammation (Evans et al., 2008). According to the European Society for Clinical Nutrition and Metabolism, cachexia is defined as "chronic disease-related malnutrition with inflammation" (Cederholm et al., 2017). Meanwhile, an international consensus on cancer cachexia in 2011 broadly classified this phenomenon into three phases: precachexia, cachexia, and refractory cachexia (Fearon et al., 2011).

Precachexia: Weight loss is minimal (i.e., 2%–5%), but there are early clinical and metabolic signs that are likely to result in future weight loss. These signs and symptoms include anorexia, features of inflammation, insulin resistance, and hypogonadism.

Cachexia: Weight loss exceeding 5% of total body weight (TBW) over the preceding 6 months, or a body mass index < 20 kg/m^2 with ongoing weight loss (> 2% TBW), or depletion of muscle mass and > 2% TBW loss (Martin et al., 2015).

Refractory cachexia: A clinically resistant catabolic state characterized by poor performance status, progressive cancer, and a life expectancy not exceeding 3 months. It must be noted that there are no strict consensus diagnostic criteria for refractory cachexia.

Cachexia occurs in the majority of cancer patients during their disease and treatment. According to the literature, in 22% of cancer patients, cachexia is directly responsible for the cause of death (Warren, 1932). Cancer causes cachexia in patients by various mechanisms, viz. (Fearon et al., 2012; Argilés et al., 2018; Grossberg et al., 2010; Martin and Kubrak, 2018):

a altering the homeostatic control of energy conservation and expenditure by neurohormonal dysregulation,
b altered hypothalamic control of appetite and satiety, resulting in anorexia and reduced food intake,
c direct interference with food intake due to symptoms of disease or its treatment (e.g., pain, odynophagia, dysgeusia, vomiting, xerostomia, etc.), and
d generation of catabolic proinflammatory cytokines and eicosanoids.

As outlined in the ASCO guidelines for managing cancer cachexia, the goal of identifying and treating cancer cachexia is to improve treatment tolerability and survival while optimizing the quality of life for patients with advanced cancer (Roeland et al., 2020). The following is a brief summary of the recommendations regarding multimodality management of cancer cachexia.

1 ***Role of a trained dietitian:*** Clinicians are encouraged to refer patients with advanced cancer suffering from anorexia and/or weight loss to a registered dietitian for assessment and dietary counseling. The goals of counseling by a dietetics expert are to provide patients and caregivers with practical and safe advice for feeding, inform them regarding which high-protein, high-calorie, nutrient-dense foods they should prioritize, and advise against fad diets and other unproven or extreme diets.
2 ***Enteral tube feeding, parenteral nutrition, and dietary supplements:*** Except in a clinical trial setting, clinicians should not routinely offer enteral tube feeding or parenteral nutrition to manage cachexia in patients with advanced cancer. Only for a select group of patients with malabsorption syndromes should parenteral nutrition be considered. When death is imminent, enteral or parenteral nutrition may be stopped. Long-chain omega-3 fatty acids (eicosapentaenoic and docosapentaenoic acids), vitamins, and minerals in dietary supplements have not significantly benefited according to current

research. Instead, natural sources of such essential elements should be encouraged to be regularly included in patients' diets.

3 ***Pharmacologic interventions:*** There are currently no FDA-approved medications for the indication of cancer cachexia due to a lack of scientific evidence. The only pharmacological therapies showing a moderate degree of improvement in cancer cachexia and endorsed by the ASCO from scientific evidence are short-term therapies of a progesterone analog (Megestrol Acetate 200–600 mg/day) or a corticosteroid (dexamethasone 4 mg/day). They are believed to improve appetite and promote weight gain. Olanzapine has shown promising results in combating cancer cachexia but has not been proven in large randomized trials yet. The choice of agent and duration of pharmacological treatment in cancer cachexia should always adhere to the goal of therapy and consider the risk-versus-benefit scenario.

Exercise: Exercise has not proven beneficial for managing cancer cachexia, so it should not be routinely promoted except in the context of a clinical trial.

EOL care

EOL care is an important component of palliative care. It prioritizes improving the quality of life for individuals with terminal conditions. It provides encouragement to their families and guides them through emotional and physical difficulties during this challenging time. EOL care seeks to alleviate physical, emotional, social, and spiritual distress while honoring the individual's autonomy and personal choices.

One of the primary objectives of EOL care is to manage physical symptoms effectively. Physical symptoms, such as pain, dyspnea, and fatigue, are common during this stage. Opioids and other pain medications are frequently used to relieve pain. Patients experiencing dyspnea may find relief through repositioning and the use of low-dose opioids or benzodiazepines (Rome et al., 2011). For terminal agitation or delirium, haloperidol is frequently administered.

Advance care planning (ACP) ensures patients' preferences are honored when they can no longer communicate them. It may include creating living wills, designating healthcare proxies, and discussing care preferences. ACP helps create a less distressing situation, and the patient receives treatment closely aligned with their wishes (Perez et al., 2013; Silveira et al., 2010).

Open and compassionate communication is the core of EOL. It helps patients and families discuss their concerns, make informed choices, and prepare for the future. However, many healthcare providers find it challenging to take these steps due to an uncertain prognosis and fear of causing more distress (Brighton and Bristowe, 2016).

Assisting families before, during, and after the death of a patient is the core objective of palliative care. Families and friends face emotional, spiritual, and practical discomforts during this period. Bereavement support, including counseling and grief resources, can aid them in coping with their loss. Preparing families for what to anticipate during the final stages may also lessen their feelings of helplessness and foster meaningful interactions with their loved ones.

EOL care encompasses more than merely symptom management. It is about valuing the individual's desires, their dignity, and their preferences. Through open communication, careful ACP, and continuous support for patients and families, palliative care teams can alleviate suffering. Moreover, it promotes a reposeful and dignified EOL experience.

Conclusion

Patients with terminal illness experience a broad spectrum of physical, emotional, social, and spiritual challenges throughout the disease process. Palliative care enhances patients' well-being by providing comprehensive, personalized support to meet their unique challenges and preferences. It plays a vital role in improving the quality of life for patients and caregivers. Palliative care encompasses the importance of interdisciplinary collaboration and patient-centered approaches to address the unique needs of patients with a terminal illness. Pain is one of the most distressing symptoms experienced by patients, significantly deteriorating their quality of life. The WHO's three-step analgesic ladder provides a structured approach to managing cancer pain. Nonpharmacological strategies frequently complement pharmacological treatments, including radiation therapy, psychological interventions, etc. Symptoms like nausea and vomiting are commonly experienced by cancer patients, particularly as side effects of chemotherapy, radiotherapy, or advanced disease stages. The causes of nausea and vomiting in oncology patients are multifactorial. Chemotherapy is the most common trigger for nausea and vomiting. CINV is classified into distinct types, with each subtype requiring specific management strategies to control nausea and vomiting. International antiemetic guidelines are available, and specific recommendations are made for selecting antiemetic agents for CINV prevention. Fatigue is one of the most common symptoms of cancer and can be a side effect of cancer treatment. It is often present before treatment and may worsen with radiation, chemotherapy, hormonal, or biological therapies. Effective management of fatigue requires a comprehensive and evidence-based approach. Integrating exercise, psychosocial interventions, and pharmacological therapies may help overcome fatigue. Dyspnea or breathlessness results from interactions among multiple physiological, psychological, social, and environmental factors and may induce secondary physiological and behavioral responses. Management of dyspnea begins with careful history taking and physical examination of the patient. MPE, PTE, SVCO, and lymphangitis carcinomatosis are specific conditions leading to dyspnea in cancer patients. Primary pharmacological measures to treat dyspnea include opioids, benzodiazepines, diuretics, and oxygen inhalation.

Cancer patients are prone to anxiety and depression from diagnosis to cure or death. International guidelines recommend that every cancer patient should be screened for anxiety at diagnosis and periodically thereafter. Management approaches include psychotherapy, psychopharmacologic treatments, and complementary treatments.

Cachexia is considered a metabolic syndrome caused by malnutrition and inflammation. Cachexia occurs in the majority of cancer patients during their disease and treatment. Cancer causes cachexia in patients through various mechanisms. Management of cancer cachexia involves a trained dietitian, adequate enteral/parenteral nutrition, drugs like megestrol acetate and dexamethasone, and exercise.

EOL care focuses on improving the quality of life for patients with serious, life-limiting illnesses. It also offers support to their families, helping them navigate the emotional and practical challenges. Open and compassionate communication is at the heart of quality EOL care.

In conclusion, palliative care is an integral, important, and indispensable component of management for patients with terminal illness.

References

Aapro, M., 2018. CINV: still troubling patients after all these years. Support Care Cancer. 26 (Suppl 1), 5–9.

Adel, N., 2017. Overview of chemotherapy-induced nausea and vomiting and evidence-based therapies. Am. J. Manag. Care 23 (14 Suppl), S259–S265.

Akechi, T., Okuyama, T., Onishi, J., Morita, T., Furukawa, T.A., 2008. Psychotherapy for depression among incurable cancer patients. Cochrane Database Syst. Rev. Published 2008 Apr 16. 2008 (2), CD005537. doi:10.1002/14651858.CD005537.pub2.

American Psychiatric Association, 2013. Diagnostic and Statistical Manual of Mental Disorders, fifth ed. DSM-5, American Psychiatric Press.

Andersen, B.L., DeRubeis, R.J., Berman, B.S., et al, 2014. Screening, assessment, and care of anxiety and depressive symptoms in adults with cancer: an American Society of Clinical Oncology guideline adaptation. J. Clin. Oncol. 32 (15), 1605–1619. doi:10.1200/JCO.2013.52.4611.

Andreas, M., Ernst, M., Kusch, M., Rußer, J.U., Csenar, M., Cryns, N., et al, 2023. Pharmacological interventions to treat adults with cancer-related fatigue (Protocol). Cochrane Database Syst. Rev. 2023 (2), CD015118.

Andrews, P.L., Sanger, G.J., 2002. Abdominal vagal afferent neurones: an important target for the treatment of gastrointestinal dysfunction. Curr. Opin. Pharmacol. 2 (6), 650–656.

Ang, S.K., Shoemaker, L.K., Davis, M.P., 2010. Nausea and vomiting in advanced cancer. Am. J. Hosp. Palliat. Care 27 (3), 219–225.

Anon, 2019. UpToDate [Internet]. Uptodate.com. Available from: https://www.uptodate.com/contents/overview-of-anxiety-in-palliative-care

Argilés, J.M., Stemmler, B., López-Soriano, F.J., et al, 2018. Inter-tissue communication in cancer cachexia. Nat. Rev. Endocrinol. 15, 9–20.

Armes, J., Chalder, T., Addington-Hall, J., Richardson, A., Hotopf, M., 2007. A randomized controlled trial to evaluate the effectiveness of a brief, behaviorally oriented intervention for cancer-related fatigue. Cancer 110, 1385–1395.

Aydin, Y., Turkyilmaz, A., Intepe, Y.S., Eroglu, A., 2009. Malignant pleural effusions: appropriate treatment approaches. Eurasian J. Med. 41 (3), 186–193. PMID: 25610100; PMCID: PMC4261269.

Bauml, J., Haas, A., CB II, S., et al, 2016. Acupuncture for dyspnea in lung cancer: results of a feasibility trial. Integr. Cancer Ther. 15 (3), 326–332.

Bogenschutz, M.P., Ross, S., Bhatt, S., et al, 2022. Percentage of heavy drinking days following psilocybin-assisted psychotherapy vs placebo in the treatment of adult patients with alcohol use disorder: a randomized clinical trial. JAMA Psychiatry. 79 (10), 953–962. [published correction appears in JAMA Psychiatry. doi: 10.1001/jamapsychiatry.2022.2982.]. doi:10.1001/jamapsychiatry.2022.2096.

Bower, J.E., 2014. Cancer-related fatigue: mechanisms, risk factors, and treatments. Nat. Rev. Clin. Oncol. 11 (10), 597–609. doi:10.1038/nrclinonc.2014.127.

Breitbart, W., Poppito, S., Rosenfeld, B., et al, 2012. Pilot randomized controlled trial of individual meaning-centered psychotherapy for patients with advanced cancer. J. Clin. Oncol. 30 (12), 1304–1309. doi:10.1200/JCO.2011.36.2517.

Brighton, L.J., Bristowe, K., 2016. Communication in palliative care: talking about the end of life, before the end of life. Postgrad. Med. J. 92 (1090), 466–470. doi:10.1136/postgradmedj-2015-133368.

Bruera, E., et al, 2013. Methylphenidate and/or a nursing telephone intervention for fatigue in patients with advanced cancer: a randomized, placebo-controlled, phase II trial. J. Clin. Oncol. 31, 2421–2427.

Carrieri, V.K., Janson-Bjerklie, S., Jacobs, S., 1984. The sensation of dyspnea: a review. Heart Lung 13 (4), 436–447. PMID: 6376427.

Cederholm, T., Barazzoni, R., Austin, P., et al, 2017. ESPEN guidelines on definitions and terminology of clinical nutrition. Clin. Nutr. 36 (1), 49–64. doi:10.1016/j.clnu.2016.09.004.

Chin-Yee, N., Tanuseputro, P., Carrier, M., Noble, S., 2019. Thromboembolic disease in palliative and end-of-life care: a narrative review. Thromb. Res. 175, 84–89. doi:10.1016/j.thromres.2018.12.028.

Chochinov, H.M., Kristjanson, L.J., Breitbart, W., et al, 2011. Effect of dignity therapy on distress and end-of-life experience in terminally ill patients: a randomised controlled trial. Lancet Oncol 12 (8), 753–762. doi:10.1016/S1470-2045(11)70153-X.

Chow, R., Simone 2nd, C.B., Rimner, A., 2024. Management of malignant superior vena cava syndrome. Ann. Palliat. Med. 13 (3), 620–626. doi:10.21037/apm-23-573.

Chu, C.C., Hsing, C.H., Shieh, J.P., Chien, C.C., Ho, C.M., Wang, J.J., 2014. The cellular mechanisms of the antiemetic action of dexamethasone and related glucocorticoids against vomiting. Eur. J. Pharmacol 722, 48–54.

Clemens, K.E., Klaschik, E., 2011. Dyspnoea associated with anxiety— symptomatic therapy with opioids in combination with lorazepam and its effect on ventilation in palliative care patients. Support Care Cancer 19 (12), 2027–2033.

Clemens, K.E., Quednau, I., Klaschik, E., 2009. Use of oxygen and opioids in the palliation of dyspnoea in hypoxic and non-hypoxic palliative care patients: a prospective study. Support Care Cancer 17 (4), 367–377.

Cruciani, R.A., et al, 2012. L-carnitine supplementation for the management of fatigue in patients with cancer: an eastern cooperative oncology group phase III, randomized, double-blind, placebo-controlled trial. J. Clin. Oncol. 30, 3864–3869.

Davis, M.P., Hallerberg, G., 2010. Palliative medicine study group of the multinational association of supportive care in C. A systematic review of the treatment of nausea and/or vomiting in cancer unrelated to chemotherapy or radiation. J. Pain Symptom Manage. 39 (4), 756–767. 14 Page 8 of 10 Curr. Treat. Options in Oncol. (2020) 21:14.

de Boer-dennert, M., de Wit, R., Schmitz, P.I., et al, 1997. Patient perceptions of the side-effects of chemotherapy: the influence of 5HT3 antagonists. Br. J. Cancer 76 (8), 1055–1061.

Delgado-Guay, M.O., Parsons, H.A., Li, Z., Palmer, L.J., Bruera, E., 2009. Symptom distress, interventions, and outcomes of intensive care unit cancer patients referred to a palliative care consult team. Cancer 115, 437–445.

Detterbeck, F.C., Jones, D.R., Morris, D.E., 2001. Palliative treatment of lung cancer. In: Detterbeck, F.C., Rivera, M.P., Socinski, M.A., Rosenman, J.G., Eds., Diagnosis and Treatment of Lung Cancer. WB Saunders Company, Philadelphia, pp. 428–433.

Doggrell, S.A., 2017. Granisetron in the treatment of chemotherapy-induced nausea and vomiting (CINV)—is there still a role after comparison with palonosetron? Expert Opin. Pharmacother. 18 (10), 1019–1026.

Dranitsaris, G., Molassiotis, A., Clemons, M., et al, 2017. The development of a prediction tool to identify cancer patients at high risk for chemotherapy-induced nausea and vomiting. Ann. Oncol. 28 (6), 1260–1267.

Dyspnea, 1999. Mechanisms, assessment, and management: a consensus statement. American Thoracic Society. Am. J. Respir. Crit. Care Med. 159 (1), 321–340. doi:10.1164/ajrccm.159.1.ats898.PMID: 9872857.

Evans, W.J., Morley, J.E., Argilés, J., et al, 2008. Cachexia: a new definition. Clin. Nutr. 27, 793–799.

Fallon, M., Giusti, R., Aielli, F., Hoskin, P., Rolke, R., Sharma, M., Ripamonti, C.I., 2018. ESMO guidelines committee. Management of cancer pain in adult patients: ESMO clinical practice guidelines. Ann. Oncol. 29 (Suppl 4), iv166–iv191.

Fearon, K.C., Glass, D.J., Guttridge, D.C., 2012. Cancer cachexia: mediators, signaling, and metabolic pathways. Cell Metab 16, 153–166.

Fearon, K., Strasser, F., Anker, S.D., et al, 2011. Definition and classification of cancer cachexia: an international consensus. Lancet Oncol 12, 489–495.

Fry, W.A., Khandekar, J.D., 1993. Parietal pleurectomy for malignant pleural effusion. Ann. Surg. Oncol. 160–164. doi:10.1007/BF02303632.

Gallagher, L.M., Lagman, R., Rybicki, L., 2018. Outcomes of music therapy interventions on symptom management in palliative medicine patients. Am. J. Hosp. Palliat. Care 35 (2), 250–257.

Glare, P., Miller, J., Nikolova, T., et al, 2011. Treating nausea and vomiting in palliative care: a review. Clin. Interv. Aging 6, 243–259.

Glare, P., Pereira, G., Kristjanson, L.J., Stockler, M., Tattersall, M., 2004. Systematic review of the efficacy of antiemetics in the treatment of nausea in patients with far-advanced cancer. Support Care Cancer 12 (6), 432–440.

Goedendorp, M.M., Gielissen, M.F., Verhagen, C.A., Bleijenberg, G., 2009. Psychosocial interventions for reducing fatigue during cancer treatment in adults. Cochrane Database Syst. Rev. 2009 (1), CD006953. doi:10.1002/14651858.CD006953.pub2.

Gontijo Garcia, G.S., Meira, K.C., de Souza, A.H., et al., 2023. Anxiety and depression disorders in oncological patients under palliative care at a hospital service: a cross-sectional study. BMC Palliat. Care 22, 116. doi:10.1186/s12904-023-01233-1.

Greer, J.A., Traeger, L., Bemis, H., et al, 2012. A pilot randomized controlled trial of brief cognitive-behavioral therapy for anxiety in patients with terminal cancer. Oncologist 17 (10), 1337–1345. doi:10.1634/theoncologist.2012-0041.

Grogono, J.C., Butler, C., Izadi, H., Moosavi, S.H., 2018. Inhaled furosemide for relief of air hunger versus sense of breathing effort: a randomized controlled trial. Respir. Res. 19 (1), 181.

Grossberg, A.J., Scarlett, J.M., Marks, D.L., 2010. Hypothalamic mechanisms in cachexia. Physiol. Behav. 100, 478–489.

Guerrero-Torrelles, M., Monforte-Royo, C., Rodríguez-Prat, A., Porta-Sales, J., Balaguer, A., 2017. Understanding meaning in life interventions in patients with advanced disease: a systematic review and realist synthesis. Palliat. Med. 31 (9), 798–813.

Gupta, M., Davis, M., LeGrand, S., Walsh, D., Lagman, R., 2013. Nausea and vomiting in advanced cancer: the Cleveland Clinic protocol. J. Support Oncol. 11, 8–13.

Hardy, J., Davis, M.P., 2021. The management of nausea and vomiting not related to anticancer therapy in patients with cancer. Curr. Treat. Options Oncol. 22 (2), 17.

Haywood, A., Duc, J., Good, P., Khan, S., Rickett, K., Vayne-Bossert, P., Hardy, J.R., 2019. Systemic corticosteroids for the management of cancer-related breathlessness (dyspnoea) in adults. Cochrane Database Syst. Rev. 2019 (2), CD012704. doi:10.1002/14651858.CD012704.pub2.

Herrstedt, J., Muss, H.B., Warr, D.G., et al, 2005. Efficacy and tolerability of aprepitant for the prevention of chemotherapy-induced nausea and emesis over multiple cycles of moderately emetogenic chemotherapy. Cancer 104, 1548–1555.

Hesketh, P.J., 2008. Chemotherapy-induced nausea and vomiting. N. Engl. J. Med. 358 (23), 2482–2494.

Hui, D., Xu, A., Frisbee-Hume, S., et al., 2014. Effects of prophylactic subcutaneous fentanyl on exercise-induced breakthrough dyspnea in cancer patients: a preliminary double-blind, randomized, controlled trial. J. Pain Symptom Manage. 47 (2), 209–217.

Jacobsen, P.B., et al, 1999. Fatigue in women receiving adjuvant chemotherapy for breast cancer: characteristics, course, and correlates. J. Pain Symptom Manage. 18, 233–242.

Jacobsen, P.B., Donovan, K.A., Vadaparampil, S.T., Small, B.J., 2007. Systematic review and meta-analysis of psychological and activity-based interventions for cancer-related fatigue. Health Psychol. 26, 660–667.

Janelsins, M.C., Tejani, M.A., Kamen, C., Peoples, A.R., Mustian, K.M., Morrow, G.R., 2013. Current pharmacotherapy for chemotherapy-induced nausea and vomiting in cancer patients. Expert Opin. Pharmacother. 14 (6), 757–766.

Jennings, A.L., Davies, A.N., Higgins, J.P., Gibbs, J.S., Broadley, K.E., 2002. A systematic review of the use of opioids in the management of dyspnoea. Thorax 57 (11), 939–944. doi:10.1136/thorax.57.11.939.

Kamen, C., Tejani, M.A., Chandwani, K., Janelsins, M., Peoples, A.R., Roscoe, J.A., Morrow, G.R., 2014. Anticipatory nausea and vomiting due to chemotherapy. Eur. J. Pharmacol. 722, 172–179.

Kangas, M., Bovbjerg, D.H., Montgomery, G.H., 2008. Cancer-related fatigue: a systematic and meta-analytic review of non-pharmacological therapies for cancer patients. Psychol. Bull. 134, 700–741.

Keeley, P.W., 2009. Nausea and vomiting in people with cancer and other chronic diseases. BMJ Clin. Evid. 2009, 2406.

Kissane, D.W., Grabsch, B., Clarke, D.M., et al, 2007. Supportive-expressive group therapy for women with metastatic breast cancer: survival and psychosocial outcome from a randomized controlled trial. Psychooncology 16 (4), 277–286. doi:10.1002/pon.1185.

Kolbrich, E.A., Goodwin, R.S., Gorelick, D.A., Hayes, R.J., Stein, E.A., Huestis, M.A., 2008. Physiological and subjective responses to controlled oral 3,4-methylenedioxymethamphetamine administration. J. Clin. Psychopharmacol. 28 (4), 432–440. doi:10.1097/JCP.0b013e31817ef470.

Kolva, E., Rosenfeld, B., Pessin, H., Breitbart, W., Brescia, R., 2011. Anxiety in terminally ill cancer patients. J. Pain Symptom Manage. 42 (5), 691–701. doi:10.1016/j.jpainsymman.2011.01.013.

Kumar, A.K., Mantri, S.N., 2025. Lymphangitic carcinomatosis. In: StatPearls [Internet]. StatPearls Publishing, Treasure Island (FL). Available from: https://pubmed.ncbi.nlm.nih.gov/32809756/. [Updated 2023 Jul 4].

LeBlanc, T.W., Abernethy, A.P., 2014. Building the palliative care evidence base: lessons from a randomized controlled trial of oxygen vs room air for refractory dyspnea. J. Natl. Compr. Canc. Netw. 12 (7), 989–992.

Lee, A.Y., Levine, M.N., Baker, R.I., et al, 2003a. Low-molecular-weight heparin versus a coumarin for the prevention of recurrent venous thromboembolism in patients with cancer. N. Engl. J. Med. 349 (2), 146–153. doi:10.1056/NEJMoa025313.

Lee, L.H., Nagarajan, C., Tan, C.W., Ng, H.J., 2021. Epidemiology of cancer-associated thrombosis in Asia: a systematic review. Front. Cardiovasc. Med. 8, 669288. Published 2021 May 21. doi:10.3389/fcvm.2021.669288.

Lee, Y.C.G., Baumann, M.H., Maskell, N.A., et al 2003b. Pleurodesis practice for malignant pleural effusions in five English-speaking countries. Chest 124, 2229–2238. doi:10.1378/chest.124.6.2229.

Light, R.W., 1995. Pleural Diseases, third ed. Williams & Wilkins, Baltimore, p. 94–116.

Lo, C., Hales, S., Jung, J., et al, 2014. Managing cancer and living meaningfully (CALM): phase 2 trial of a brief individual psychotherapy for patients with advanced cancer. Palliat. Med. 28 (3), 234–242. doi:10.1177/0269216313507757.

Lorusso, D., Bria, E., Costantini, A., Di Maio, M., Rosti, G., Mancuso, A., 2017. Patients' perception of chemotherapy side effects: expectations, doctor-patient communication and impact on quality of life—an Italian survey. Eur. J. Cancer Care (Engl.) 26 (2), e12618.

Lutz, S., Balboni, T., Jones, J., Lo, S., Petit, J., Rich, S.E., Wong, R., Hahn, C., 2017. Palliative radiation therapy for bone metastases: update of an ASTRO evidence-based guideline. Pract. Radiat. Oncol. 7 (1), 4–12.

Lutz, S., Berk, L., Chang, E., Chow, E., Hahn, C., Hoskin, P., Howell, D., Konski, A., Kachnic, L., Lo, S., Sahgal, A., Silverman, L., von Gunten, C., Mendel, E., Vassil, A., Bruner, D.W., Hartsell, W., 2011. American society for radiation oncology (ASTRO). Palliative radiotherapy for bone metastases: an ASTRO evidence-based guideline. Int. J. Radiat. Oncol. Biol. Phys. 79 (4), 965–976.

Maeda, T., Hayakawa, T., 2017. Range of effective corticosteroid doses for alleviating dyspnea in terminal cancer patients: a retrospective review. J. Pain Palliat. Care Pharmacother. 31 (1), 10–15.

Martin, L., Kubrak, C., 2018. How much does reduced food intake contribute to cancer-associated weight loss? Curr. Opin. Support Palliat. Care 12, 410–419.

Martin, L., Senesse, P., Gioulbasanis, I., et al, 2015. Diagnostic criteria for the classification of cancer-associated weight loss. J. Clin. Oncol. 33, 90–99.

Mestdagh, F., Steyaert, A., Lavand'homme, P., 2023. Cancer pain management: a narrative review of current concepts, strategies, and techniques. Curr. Oncol. 30 (7), 6838–6858.

Minchom, A., Punwani, R., Filshie, J., et al, 2016. A randomized study comparing the effectiveness of acupuncture or morphine versus the combination for the relief of dyspnoea in patients with advanced non-small cell lung cancer and mesothelioma. Eur. J. Cancer 61, 102–110.

Minton, O., Richardson, A., Sharpe, M., Hotopf, M., Stone, P., 2014. A systematic review and meta-analysis of the pharmacological treatment of cancer-related fatigue. Br. J. Cancer 111 (4), 574–581.

Moraska, A.R., et al, 2010. Phase III, randomized, double-blind, placebo-controlled study of long-acting methylphenidate for cancer-related fatigue: north Central Cancer Treatment Group NCCTG-N05C7 trial. J. Clin. Oncol. 28, 3673–3679.

Moyer, A., Sohl, S.J., Knapp-Oliver, S.K., Schneider, S., 2009. Characteristics and methodological quality of 25 years of research investigating psychosocial interventions for cancer patients. Cancer Treat. Rev. 35, 475–484.

National Comprehensive Cancer Network (NCCN), 2024a. Antiemesis (Version 1.2024) [Internet]. NCCN Clinical Practice Guidelines in Oncology. Available from: https://www.nccn.org/guidelines/guidelinesdetail?category=3&id=1415., [cited 2024 Mar 18].

National Comprehensive Cancer Network (NCCN), 2024b. Cancer-Related Fatigue (Version 1.2024) [Internet]. NCCN Clinical Practice Guidelines in Oncology. Available from: https://www.nccn.org/guidelines/guidelines-detail?category=3&id=1424. [cited 2024 Mar 18].

Navari, R.M., 2014. Olanzapine for the prevention and treatment of chronic nausea and chemotherapy-induced nausea and vomiting. Eur. J. Pharmacol 722, 180–186.

Navari, R.M., Nagy, C.K., Gray, S.E., 2013. The use of olanzapine versus metoclopramide for the treatment of breakthrough chemotherapy-induced nausea and vomiting in patients receiving highly emetogenic chemotherapy. Support Care Cancer 21, 1655–1663.

Navigante, A.H., Cerchietti, L.C., Castro, M.A., Lutteral, M.A., Cabalar, M.E., 2006. Midazolam as an adjunct therapy to morphine in the alleviation of severe dyspnea perception in patients with advanced cancer. J. Pain Symptom Manage. 31 (1), 38–47.

Nicholson, A.A., Ettles, D.F., Arnold, A., Greenstone, M., Dyet, J.F., 1997. Treatment of malignant superior vena cava obstruction: metal stents or radiation therapy. J. Vasc. Interv. Radiol. 8 (5), 781–788. doi:10.1016/s1051-0443(97)70660-2.

Paget, S., 1989. The distribution of secondary growths in cancer of the breast. Cancer Metastasis Rev. 8, 98–101.

Pearson, E.J.M., Morris, M.E., Di Stefano, M., McKinstry, C.E., 2018. Interventions for cancer-related fatigue: a scoping review. Eur. J. Cancer Care (Engl.) 27 (1). doi:10.1111/ecc.12516.

Perez, M.D.V., Macchi, M.J., Agranatti, A.F., 2013. Advance directives in the context of end-of-life palliative care. Curr. Opin. Support Palliat. Care 7 (4), 406–410. doi:10.1097/SPC.0000000000000007.

Phillips, K.M., Pinilla-Ibarz, J., Sotomayor, E., Lee, M.R., Jim, H.S., Small, B.J., Sokol, L., Lancet, J., Tinsley, S., Sweet, K., Komrokji, R., Jacobsen, P.B., 2013. Quality of life outcomes in patients with chronic myeloid leukemia treated with tyrosine kinase inhibitors: a controlled comparison. Support Care Cancer 21 (4), 1097–1103.

Puetz, T.W., Herring, M.P., 2012. Differential effects of exercise on cancer-related fatigue during and following treatment: a meta-analysis. Am. J. Prev. Med. 43, e1–e24.

Puspawati, N.L.P.D., Sitorus, R., Herawati, T., 2017. Hand-held fan airflow stimulation relieves dyspnea in lung cancer patients. Asia Pac. J. Oncol. Nurs. 4 (2), 162–167.

Putnam, J.B., Light, R.W., Rodriguez, R.M., et al, 1999. A randomized comparison of indwelling pleural catheter and doxycycline pleurodesis in the management of malignant pleural effusions. Cancer 86, 1992–1999.

Rao, K.V., Faso, A., 2012. Chemotherapy-induced nausea and vomiting: optimizing prevention and management. Am. Health Drug Benefits 5 (4), 232–240.

Rich, S.E., et al, 2018. Update of the systematic review of palliative radiation therapy fractionation for bone metastases. Radiother. Oncol. 126 (3), 547–557. doi:10.1016/j.radonc.2018.01.003.

Rodriguez-Panadero, F., Borderas Naranjo, F., Lopez-Mejias, J., 1989. Pleural metastatic tumours and effusions: frequency and pathogenic mechanisms in a post-mortem series. Eur. Respir. J. 2, 366–369.

Roeland, E.J., Bohlke, K., Baracos, V.E., et al, 2020. Management of cancer cachexia: ASCO guideline. J. Clin. Oncol. 38 (21), 2438–2453. doi:10.1200/JCO.20.00611.

Rogers, J.B., Modi, P., Minteer, J.F., 2025. Dyspnea in palliative care. In: StatPearls [Internet]. StatPearls Publishing, Treasure Island (FL). Available from: https://pubmed.ncbi.nlm.nih.gov/30252378/. [Updated 2023 Apr 3].

Roila, F., Molassiotis, A., Herrstedt, J., et al, 2016. MASCC and ESMO guideline update for the prevention of chemotherapy- and radiotherapy-induced nausea and vomiting and of nausea and vomiting in advanced cancer patients. Ann. Oncol. 27 (suppl 5), v119–v133.

Rojas, C., Raje, M., Tsukamoto, T., Slusher, B.S., 2014. Molecular mechanisms of 5-HT(3) and NK(1) receptor antagonists in prevention of emesis. Eur. J. Pharmacol. 722, 26–37.

Rome, R.B., Luminais, H.H., Bourgeois, D.A., Blais, C.M., 2011. The role of palliative care at the end of life. Ochsner J. 11 (4), 348–352.

Ryan, R., Spathis, A., Booth, S., 2014. Correlates between basic science and therapeutic interventions: the theory and the practice. Curr. Opin. Support Palliat. Care 8 (3), 200–207.

Shankar, A., Roy, S., Malik, A., Julka, P.K., Rath, G.K., 2015. Prevention of chemotherapy-induced nausea and vomiting in cancer patients. Asian Pac. J. Cancer Prev. 16 (15), 6207–6213.

Silveira, M.J., Kim, S.Y.H., Langa, K.M., 2010. Advance directives and outcomes of surrogate decision making before death. N. Engl. J. Med. 362 (13), 1211–1218. doi:10.1056/NEJMsa0907901.

Smith, H.S., Smith, J.M., Smith, A.R., 2012. An overviewof nausea vomiting in palliative medicine. Ann. Palliat. Med. 1, 103–114.

Snijders, R.A.H., Brom, L., Theunissen, M., van den Beuken-van Everdingen, M.H.J., 2023. Update on prevalence of pain in patients with cancer 2022: a systematic literature review and meta-analysis. Cancers 15, 591.

Spitzer, R.L., Kroenke, K., Williams, J.B., Löwe, B., 2006. A brief measure for assessing generalized anxiety disorder: the GAD-7. Arch. Intern. Med. 166 (10), 1092–1097. doi:10.1001/archinte.166.10.1092.

Storey, P., Knight, C.F., 1997. UNIPAC Two: Alleviating Psychological and Spiritual Pain in the Terminally Ill. Hospice/Palliative Care Training for Physicians: A Self-Study Program. American Academy of Hospice and Palliative Medicine, Gainesville, FL.

Strasser, B., Steindorf, K., Wiskemann, J., Ulrich, C.M., 2013. Impact of resistance training in cancer survivors: a meta-analysis. Med. Sci. Sports Exerc. 45, 2080–2090.

Swan, F., Newey, A., Bland, M., et al, 2019. Airflow relieves chronic breathlessness in people with advanced disease: an exploratory systematic review and meta-analyses. Palliat. Med. 33 (6), 618–633.

Tan, L., Liu, J., Liu, X., et al, 2009. Clinical research of Olanzapine for prevention of chemotherapy-induced nausea and vomiting. J. Exp. Clin. Cancer Res. 28, 131.

Temel, J.S., Greer, J.A., Muzikansky, A., Gallagher, E.R., Admane, S., Jackson, V.A., et al, 2010. Early palliative care for patients with metastatic non–small-cell lung cancer. N. Engl. J. Med. 363 (8), 733–742.

Thomas, E.M., Weiss, S.M., 2010. Nonpharmacological interventions with chronic cancer pain in adults. Cancer Control 7 (2), 157–164.

Tremblay, P.B., Kaiser, R., Sezer, O., Rosler, N., Schelenz, C., Possinge, K., et al, 2003. Variations in the 5-hydroxytryptamine type 3B receptor gene as predictors of the efficacy of antiemetic treatment in cancer patients. J. Clin. Oncol. 21 (11), 2147–2155.

Warren, S., 1932. The immediate cause of death in cancer. Am. J. Med. Sci. 184, 610–613.

WHO, 2018. Guidelines for the Pharmacological and Radiotherapeutic Management of Cancer Pain in Adults and Adolescents. World Health Organization, Geneva.

World Health Organization. 2025. Palliative care [Internet]. Available from: https://www.who.int/news-room/fact-sheets/detail/palliative-care. [cited 2025 Jan 22].

Yamamoto, Y., Watanabe, H., Sakurai, A., et al, 2018. Effect of continuous intravenous oxycodone infusion in opioid-naive cancer patients with dyspnea. Jpn. J. Clin. Oncol. 48 (8), 748–752.

Yates, P., et al, 2005. Randomized controlled trial of an educational intervention for managing fatigue in women receiving adjuvant chemotherapy for early-stage breast cancer. J. Clin. Oncol. 23, 6027–6036. PubMed: 16135471.

Yates, P., Schofield, P., Zhao, I., Currow, D., 2013. Supportive and palliative care for lung cancer patients. J. Thorac. Dis. 5 (Suppl 5), S623–S628.

Zemel, R.A., 2022. Pharmacologic and non-pharmacologic dyspnea management in advanced cancer patients. Am. J. Hosp. Palliat. Care 39 (7), 847 855. doi:10.1177/10499091211040436.

CHAPTER

8

Adverse effects of cancer pharmacotherapy: Prevention and management

Rituparna Chetia[a] *and Manas Dubey*[b,c]

[a]Department of Medical Oncology/Hematology, Medicine, Clinical Hematology, AIIMS, Guwahati, India
[b]Medical Oncology, Tata Memorial Centre, Guwahati, India
[c]Consultant Medical and Hemato-Oncologist, Kamla Nehru Memorial Hospital, Prayagraj, India

Infections in cancer

Patients with malignancies are at an increased risk of infection due to the disease and its treatment. There are mainly two categories of risk factors: cancer-related and treatment-related factors.

The cancer/malignancy-related risk factors are further subdivided into hematologic and solid tumors.

Malignancy-related factors

The following hematologic malignancies are associated with infections:

1. Myelodysplastic syndrome (MDS)/acute leukemia, due to prolonged neutropenia
2. Chronic lymphocytic leukemia (CLL)/multiple myeloma, due to hypogammaglobulinemia. There is also impaired B-cell immunity in multiple myeloma
3. Acute lymphoblastic leukemia, due to steroid therapy
4. Adult T-cell leukemia/lymphoma, hairy cell leukemia/Hodgkin lymphoma, due to defective T-cell immunity

https://doi.org/10.1016/B978-0-443-38307-6.00006-5

Solid tumors prone to infection are:

1. Hepatobiliary tumors
2. Head and neck cancers
3. Genitourinary cancers
4. Colon cancer
5. Endobronchial tumors

Treatment-related factors

1. Mucositis, prone to candidiasis and bacterial infections
2. Neutropenia, prone to bacterial and fungal infections
3. Corticosteroids, prone to bacterial, mold, and viral infections
4. Rituximab, prone to bacterial infections, *Pneumocystis jirovecii*, and varicella-zoster virus (VZV) infections
5. Nucleoside analogs, prone to bacterial infections, *P. jirovecii*, and *Candida. neoformans*
6. Alemtuzumab, prone to bacteria, viruses, *P. jirovecii*, and fungi
7. Bruton tyrosine kinase inhibitors, prone to aspergillosis, *P. jirovecii*, and *C. neoformans* infections
8. Calcineurin inhibitors, prone to VZV and *P. jirovecii* infections
9. Radiation therapy, prone to bacterial infections, mucosal candidiasis, and HSV infection

Prevention of infections

Infection in patients having malignancy can be prevented by avoidance of exposure, chemoprophylaxis, and immunization.

Prevention of bacterial infections

Prophylaxis against pneumococcal infection

Pneumococcal prophylaxis should be considered for individuals who are asplenic or for allogeneic Hematopoietic Stem Cell Transplant (HSCT) recipients with chronic graft-versus-host disease (GVHD). Among these transplant recipients, pneumococcal infections generally occur in the later stages posttransplant. Both vaccination and penicillin prophylaxis are advised for such patients.

Patients who have undergone splenectomy or have functional hyposplenism should receive immunization against *Haemophilus influenzae* type b, *Streptococcus pneumoniae*, and *Neisseria meningitidis*, along with an annual influenza vaccine. Ideally, these vaccines should be given either 2 weeks before or 2 weeks after a splenectomy, but they may be administered at any point before or after the procedure.

Since current vaccines do not provide protection against all strains of pneumococcus, *H. influenzae*, and meningococcus, daily antibiotic prophylaxis is essential for high-risk hyposplenic patients (Ladhani et al., 2024).

Antibacterial prophylaxis in afebrile neutropenic patients

Fluoroquinolone prophylaxis is recommended for patients at high risk of febrile neutropenia (FN) or severe, prolonged neutropenia. This includes most individuals with acute myeloid leukemia (AML), MDSs, or those undergoing hematopoietic stem cell transplantation (HSCT) with myeloablative conditioning regimens.

Profound neutropenia is defined as an absolute neutrophil count (ANC) below 100/μL (0.1 × 10^9/L). Neutropenia is considered prolonged if it persists for 7 days or more. Routine antibiotic prophylaxis is not recommended for patients with solid tumors (Flowers et al., 2013).

Prevention of fungal infections

The primary fungal pathogens affecting cancer patients are *Candida* species and molds. Key risk factors for fungal infections include broad-spectrum antibiotic use, central venous catheters, chemotherapy-induced mucositis, total parenteral nutrition, prolonged ICU stays, and neutropenia.

The risk of invasive mold infections is closely linked to the duration of neutropenia and prolonged steroid use for GVHD in allogeneic hematopoietic stem cell transplant (HSCT) recipients. The most common mold infections are aspergillosis and mucormycosis.

Antifungal prophylaxis with an oral triazole or a parenteral echinocandin is recommended for patients at high risk of severe, prolonged neutropenia, including most individuals with AML, MDSs, or those undergoing HSCT. Routine antifungal prophylaxis is not recommended for patients with solid tumors (Flowers et al., 2013).

Prophylaxis of viral infections

Antiviral prophylaxis against herpes simplex virus (HSV) is recommended for patients undergoing chemotherapy for acute leukemia, all HSV-seropositive allogeneic hematopoietic stem cell transplant (HSCT) recipients, and select autologous HSCT recipients at high risk of mucositis during the neutropenic phase. Infections caused by HSV and VZV are frequently observed in patients receiving alemtuzumab, as well as proteasome inhibitors such as bortezomib and lenalidomide (Swaika et al., 2012).

Cytomegalovirus (CMV) disease is rare in most cancer patients but is a significant pathogen following allogeneic hematopoietic stem cell transplantation (HSCT). CMV-seronegative HSCT recipients receiving grafts from CMV-seronegative donors should receive transfusion products exclusively from CMV-negative donors. If this is not possible, leukocyte-depleted blood products should be used as an alternative (Nichols et al., 2003).

Prevention of *P. jirovecii* pneumonia

Pneumocystis Carinii Pneumonia (PCP) prophylaxis should be considered for cancer patients receiving high-dose corticosteroids, individuals with CLL treated with idelalisib or fludarabine, and allogeneic HSCT recipients until at least day 180 posttransplant, or longer if immunosuppressive therapy continues. Prophylaxis is also recommended for patients receiving alemtuzumab, as well as those with gliomas undergoing treatment with temozolomide and radiation or corticosteroids. The most effective agent for PCP prophylaxis is trimethoprim/sulfamethoxazole (Cooley et al., 2014).

Fungal infections in patients with cancer

Most common infections are caused by *Candida* and molds.

Candida causes oropharyngeal and esophageal candidiasis, as well as candidemia. The most common *Candida* species is *Candida albicans*, and it is usually susceptible to fluconazole. As per guidelines, all intravascular catheters should be removed in patients who develop candidemia. Intensive chemotherapy as used in acute leukemia can cause chronic disseminated candidiasis.

Invasive aspergillosis is commonly diagnosed in a chest CT scan and frequently encountered in neutropenic patients and HSCT recipients. Characteristic findings include well-circumscribed nodules, halo sign, and crescent sign in the CT scan (Caillot et al., 2001).

The serum *Aspergillus* galactomannan and β-D-glucan assays are immunoassays that detect fungal antigens in peripheral blood and have been accepted as diagnostic adjuncts for invasive fungal infections. Voriconazole is the drug of choice for invasive aspergillosis. Isavuconazole has been found to be noninferior to voriconazole in recent studies (Maertens et al., 2016).

Mucormycosis commonly manifests as rhinocerebral or pulmonary disease. Prolonged neutropenia, steroid therapy, iron overload, desferrioxamine therapy, diabetic ketoacidosis, and protein-calorie malnutrition are the risk factors for this fungal infection. Fever, headache, and facial pain are common presentations. Treatment involves amphotericin B plus early and aggressive surgical debridement in cases of localized disease (Roden et al., 2005).

Catheter-related infections

Infections and bacteremia due to intravenous catheters are commonly encountered in patients receiving cytotoxic chemotherapy. They can occur without any signs of localized infections. For central line-related blood stream infections, the catheter should be removed, and systemic antibiotics should be instituted. In hemodynamically stable patients with permanent catheters, salvage may be considered. Antibiotic lock therapy is sometimes effective (Mermel et al., 2009).

Skin and soft tissue infections, abdominal infections, pulmonary infections, genitourinary tract infections, etc., are other infections encountered in patients receiving chemotherapy and HSCT recipients.

Neutropenia and thrombocytopenia

Chemotherapy-induced neutropenia (CIN) is a significant contributor to hematological complications and limits the dosage of chemotherapy. This condition can lead to inadequate disease management and lower survival rates. As a result, patients may miss out on potential cures due to the severe implications of CIN. Factors associated with CIN include advanced age, compromised functional and nutritional health, the presence of serious comorbidities, the specific type of cancer, prior chemotherapy treatments, the disease's stage, particular chemotherapy protocols, and the use of combination therapies. CIN is typically defined by an ANC of <2000 cells/mm^3 in the bloodstream. The severity of CIN is further categorized using the National Cancer Institute's Common Terminology Criteria for Adverse Events, version 5.0. This grading system classifies neutropenia into four levels: (1) Grade 1 with an ANC of 1500–2000 cells/mm^3, (2) Grade 2 with an ANC of 1000–1500 cells/mm^3, (3) Grade 3 with an ANC of 500–1000 cells/mm^3, and (4) Grade 4 with an ANC below 500 cells/mm^3. A critical complication arising from CIN is FN, which can lead to heightened healthcare costs, severe infections, prolonged hospitalizations, intensive treatment, increased morbidity, and mortality.

FN is a prevalent cause of oncological emergencies. FN is defined as a fever of ≥ 38.3°C orally or ≥38.0°C lasting more than 1 h, occurring alongside neutropenia (ANC < 500 cells/μL). In patients with solid tumors, the incidence of FN ranges from 13% to 21% with standard myelosuppressive chemotherapy, and this rate is significantly higher during the first cycle, reaching 23%–36%. FN is categorized into three risk levels: high risk (greater than 20%), intermediate risk (10%–20%), and low risk (<10%). Various assessment tools, such as the Multinational Association for Supportive Care in Cancer and the Clinical Index of Stable Febrile Neutropenia (CISNE) score model, have been created to evaluate the risk of complications associated with FN. The CISNE score model is particularly beneficial for guiding treatment decisions in patients with stable FN. For primary prevention, granulocyte colony-stimulating factors (GCSFs) are advised for use during the first and subsequent cycles of chemotherapy when the risk of FN is high (over 20%). Patients identified as having a high risk of FN complications (CISNE score ≥3) should be admitted to the hospital and treated with intravenous broad-spectrum antibiotics, with initial empirical therapy aimed at addressing *Pseudomonas aeruginosa* and other serious Gram-negative bacteria. For patients at low risk of FN complications (CISNE score 0–2), oral empirical antibiotic treatment may be appropriate for select individuals. Ongoing challenges include determining the advantages of GCSFs and establishing the most effective antibiotic protocols in FN emergencies.

Cancer treatments are recognized for their potential to induce thrombocytopenia (TCP). Specifically, chemotherapy-induced TCP (CIT) refers to the decrease in platelet count resulting from chemotherapy agents. The typical platelet count range is between 150,000 and 450,000/μL.

The National Cancer Institute Common Terminology Criteria for Adverse Events (NCI CTCAE) categorizes TCP by grade: Grade 1 indicates a count of 75,000–150,000/μL, Grade 2 ranges from 50,000 to 75,000/μL, Grade 3 is between 25,000 and 50,000/μL, and Grade 4 is defined as <25,000/μL. In clinical practice, surgical interventions are generally avoided when platelet counts fall below 50,000/μL, and for central nervous system (CNS) procedures, the threshold is set at 75,000–100,000/μL.

The diagnostic approach for CIT mirrors that of TCP from other origins. All patients undergoing cancer treatment should have a complete blood count and a peripheral blood smear analysis to check for abnormal cells, such as schistocytes. In addition, metabolic tests should include liver enzyme levels, lactate dehydrogenase, and haptoglobin to exclude other causes. Spontaneous bleeding can occur when platelet counts drop below 20,000/µL, with a higher risk at levels under 5000–10,000/µL. Prophylactic platelet transfusions can help avert spontaneous bleeding.

Treatment of CIT

CIT is managed identically to other causes of TCP:

1. Delaying the scheduled delivery of a chemotherapy agent or temporarily stopping the oral targeted drug.
2. Managing the life-threatening complications, for example, platelet transfusions when platelet counts drop below 10,000/µL.
3. Dose reduction in further chemotherapy
4. Thrombopoietin receptor agonists have shown mixed results in CIT (Punekar, 2020).

Nausea and vomiting

Nausea and vomiting rank as the most prevalent side effects experienced by patients undergoing chemotherapy. Chemotherapy-induced nausea and vomiting (CINV) affects nearly 40% of cancer patients receiving treatment. It is essential for healthcare providers to recognize CINV and proactively address the symptoms to enhance patient comfort and adherence to treatment. Emetogenic drugs are classified into four categories based on their potential to induce CINV: minimal (<10%), low (10%–30%), moderate (30%–90%), and high (>90%). For instance, highly emetogenic chemotherapy regimens often involve high doses of cisplatin, carmustine, cyclophosphamide exceeding 1500 mg/m^2, dacarbazine, mechlorethamine, streptozocin, and combinations of anthracyclines and phosphamide (AC). Moderately emetogenic chemotherapy regimens can vary but may include agents such as carboplatin, doxorubicin, irinotecan, oxaliplatin, and cyclophosphamide. A variety of antiemetic medications with distinct mechanisms have been developed to combat CINV, most of which are administered as preventive treatments. Currently, the most frequently prescribed antiemetic drugs include 5-HT3 antagonists, corticosteroids, and NK1 receptor antagonists. Guidelines for antiemetic therapy are:

For high emetogenicity

On day 1

Aprepitant, 5-HT3 receptor antagonist, and dexamethasone

or

Fosaprepitant, 5-HT3 receptor antagonist, and dexamethasone

or

Rolapitant, 5-HT3 receptor antagonist, and dexamethasone

or

Netupitant plus dexamethasone

or

Olanzapine, palonosetron, dexamethasone
or
Aprepitant, 5-HT3 receptor antagonist, dexamethasone, and olanzapine

On days 2, 3, and 4

Aprepitant on days 2 and 3, and dexamethasone on days 2, 3, and 4
or
Dexamethasone once daily on day 2, then dexamethasone twice daily on days 3 and 4
or
Dexamethasone twice daily
or
Dexamethasone daily
or
Olanzapine daily
or
Aprepitant on days 2 and 3, dexamethasone on days 2, 3, and 4, and olanzapine on days 2, 3, and 4

For moderate emetogenicity

On day 1

5-HT3 receptor antagonist and dexamethasone
or
Aprepitant, 5-HT3 receptor antagonist, dexamethasone
or
Fosaprepitant, 5-HT3 receptor antagonist, dexamethasone
or
Olanzapine, palonosetron, dexamethasone

On days 2, 3, and 4

Dexamethasone or 5-HT3 receptor antagonist monotherapy
or
Aprepitant ± dexamethasone
or
± dexamethasone
or
Olanzapine

For low emetogenicity

One dose of dexamethasone, metoclopramide, and prochlorperazine.

Management of CINV remains a great challenge, with many needs waiting to be adequately addressed, such as controlling nonacute CINV, developing appropriate CINV treatment protocols for multiple-day chemotherapy patients, and providing options for those prone to CINV despite treatment. Further research is required to optimize CINV management for these patients (Gupta et al., 2021).

Diarrhea and constipation

Both diarrhea and constipation are major causes of morbidity in patients with cancer. Severe diarrhea can even be life-threatening at times. It can lead to dehydration, electrolyte imbalance, malnutrition, and pressure ulcer formation. Hence, it is very important that the underlying cause is evaluated and managed (Tarricone et al., 2016).

Diarrhea

5-Fluorouracil (5-FU), capecitabine, and irinotecan most commonly cause chemotherapy-induced diarrhea (Andreyev et al., 2014).

Neutropenic colitis

Also known as necrotizing enterocolitis or typhlitis, neutropenic enterocolitis is a life-threatening complication of chemotherapy. It presents with abdominal pain, rebound tenderness, bloody diarrhea, and fever in the context of neutropenia, typically occurring when the ANC falls below 500/μL (Cloutier, 2009).

Microbial infection leads to necrosis of various layers of the bowel wall, with multiple contributing factors, including mucosal injury from cytotoxic drugs, profound neutropenia, and impaired host defenses against invasion by microorganisms (Cloutier, 2009).

Diagnosis is based on clinical evaluation and imaging, with a CT scan being the preferred modality. Abdominal ultrasonography may also detect bowel wall thickening, often appearing as a target or halo sign.

Radiotherapy-induced diarrhea

Radiation enteritis commonly occurs following treatment for cancers of the anus, rectum, cervix, uterus, prostate, bladder, and testes, as well as after total-body irradiation.

Acute radiation enteritis develops within 6 weeks of therapy, whereas late radiation enteritis typically manifests 8–12 months after treatment. In some cases, symptoms may be delayed for several years (Theis et al., 2010).

Management of diarrhea

Patient should be rehydrated either orally or parenterally. Stool workup to rule out infectious causes of diarrhea should be done. Antidiarrheal medications such as opioids, somatostatin analogs, and budesonide can be used. In cases of radiation-induced diarrhea which is prolonged, high-dose loperamide and octreotide therapy, if required, are indicated.

The mortality rate is very high in neutropenic colitis. It should be managed with broad-spectrum antibiotics, granulocyte colony-stimulating factors, IV fluids, and bowel rest. In a few patients, surgery may be helpful to avoid progressive bowel necrosis and perforation and to help control sepsis.

Constipation

It is most commonly encountered in the opioid-treated population. Other agents causing constipation are vinca alkaloids, thalidomide, serotonin 5-HT3 receptor antagonists, iron supplements, diuretics, etc. (Haim et al., 1994).

General measures like increasing dietary fiber intake, fluid intake, and physical activity can prevent constipation. The role of laxatives is limited. Enemas or suppositories are other alternative approaches.

Oral complications

Chemotherapy and radiation therapy are the primary methods employed in cancer treatment. Patients undergoing these treatments often experience oral complications such as mucositis (stomatitis), xerostomia (dry mouth), and infections (bacterial, fungal, or viral, particularly in neutropenic individuals). Other issues may include dental caries, altered taste sensation, and osteoradionecrosis of the jaw. Oral mucositis is a significant nonhematologic side effect of chemotherapy, leading to considerable morbidity, pain, difficulty swallowing (dysphagia), and subsequent risks of dehydration and malnutrition.

Severe oral toxicities can hinder the effective administration of cancer chemotherapy regimens. For instance, adjustments in chemotherapy dosages or changes to the treatment schedule may be necessary to allow for the healing of oral lesions. In cases of severe oral complications (grades 3–4), patients may be unable to continue their cancer treatment, resulting in interruptions. Such disruptions in medication dosing can have a direct impact on oncological outcomes and overall cancer survival rates.

Key risk factors for these complications include the patient's age, nutritional status, type of cancer, oral hygiene practices during treatment, and neutrophil levels prior to therapy.

Prevention of oral mucositis

Prevention measures play important roles in the incidence of mucositis.

Oral hygiene: **-** Comprehensive oral assessment by a dentist before cancer therapy is initiated.

Cryotherapy: Cooling of the oral mucosa using ice chips reduces blood flow to the oral mucosa, thus reducing the availability of chemotherapeutic agents in the oral cavity.

Antimicrobial/antifungal/antiviral agents: The oral mucosa in cancer patients often harbors pathogenic microorganisms, including both Gram-positive and Gram-negative bacteria, as well as fungi. Antibiotic lozenges can help prevent severe cases of oral mucositis. For patients with compromised immune systems or those experiencing myelosuppression, both topical and systemic acyclovir treatments are effective for managing oral herpetic infections.

Treatment of oral mucositis

Patient education is very important in managing chemotherapy-induced mucositis. Patients are encouraged to eat slowly, have frequent meals, avoid hard or sharp foods, increase their intake of soft foods, and avoid hot drinks.

Oral care: Routine mouth care should be performed every 4 days. Patients should be counseled to rinse their mouth thoroughly after every meal so that food particles do not remain in the mouth. Antibiotic rinses, such as chlorhexidine, may also be used.

Topical agents: Sucralfate and magnesium hydroxide are some of the film-forming or coating agents that may be beneficial in treating established mucositis.

Topical anesthetics: Local anesthetics, such as lidocaine and capsaicin, can also be used.

Pulmonary toxicity

It can be caused by both chemotherapy and radiotherapy.

Chemotherapy can lead to lung toxicity through several pathways. This includes direct damage to alveolar epithelial cells, triggering an inflammatory immune response, and causing injury or activation of endothelial cells, which can result in capillary leak syndrome.

Drugs commonly implicated in causing pulmonary toxicity are bleomycin, mitomycin, methotrexate, G-CSF, IL-2, EGFR inhibitors such as gefitinib, erlotinib, and cetuximab, Kit/BCR-ABL inhibitors such as imatinib and dasatinib, mTOR inhibitors such as everolimus and sirolimus, bortezomib, thalidomide, lenalidomide, carfilzomib, all-trans-retinoic acid (ATRA), and CD19-directed CAR T-cell infusion (Inomata et al., 2004; Lin et al., 2006).

Factors such as combination chemotherapy, concurrent radiation therapy, oxygen treatment, renal issues, existing lung conditions like idiopathic pulmonary fibrosis and chronic obstructive pulmonary disease, radiation pneumonitis, widespread lung metastasis, poor overall health, and older age are linked to a higher risk of pulmonary toxicity.

Most common presenting feature of drug-induced pulmonary toxicity is dyspnea. The most frequent radiographic finding is a reticular or nodular interstitial pattern, often characterized by thickened septal lines, which can appear either basilar or diffuse. In cases of chemotherapy-related pulmonary toxicity, common abnormalities in pulmonary function tests include a decreased diffusing capacity for carbon monoxide and, in more severe instances, a restrictive ventilatory defect.

Lung biopsy is usually necessary for a diagnosis and to rule out other causes of lung injury.

The best way to handle pulmonary toxicity linked to a chemotherapeutic agent is through prevention. If toxicity occurs, the first step in treatment is to remove the chemotherapeutic agent.

Radiation induces pulmonary toxicity by producing reactive oxygen species and reactive nitrogen species in the first phase. In the second phase, inflammation in response to injury occurs, and the last phase is the process of repair. In mild cases, the changes subside entirely, but in severe cases, a chronic phase may persist or progress from months to years (Citrin et al., 2013).

Certain factors are known to increase the risk of developing classic radiation pneumonitis. These include being female, having a primary lung cancer located in the lower lobe, undergoing concurrent chemotherapy, having received prior radiation treatment, and stopping steroid medications (Graves et al., 2010).

Symptoms of acute radiation pneumonitis typically appear 2–3 months after finishing treatment. The main presenting feature of radiation pneumonitis is dyspnea. Patients may present with a nonproductive cough. The diagnosis of radiation pneumonitis can often be made based on clinical factors, including the timing of radiation in relation to the onset of symptoms and the characteristic appearance on chest X-rays, such as infiltrates that align with the edges of the irradiated area. Steroids have shown promising results in the treatment of radiation pneumonitis.

Cardiotoxicity

The cardiotoxic effects of chemotherapy present significant challenges due to the limited regenerative ability of myocardial tissue, rendering the heart vulnerable to both temporary and permanent adverse effects from chemotherapy agents. Clinically, these effects can manifest as hypertension, myocardial ischemia, arrhythmias, thromboembolism, systolic dysfunction, and congestive heart failure (CHF). Delays in diagnosing and treating cardiotoxic complications can lead to considerable cardiac morbidity and postponements in scheduled treatments. Common chemotherapy agents linked to cardiotoxicity include anthracyclines like doxorubicin, alkylating agents such as cyclophosphamide and ifosfamide, the antimicrotubular agent docetaxel, and antimetabolites like 5-FU and capecitabine. Among these, anthracyclines are the most frequently implicated and have been associated with cardiomyopathy and CHF.

Various strategies have been suggested to prevent cardiotoxicities associated with chemotherapy. One effective method is to limit the cumulative dose of anthracyclines to < 400 mg/m^2, which helps mitigate the risk of cardiotoxicity. In addition, using slow continuous intravenous infusion instead of bolus administration has been linked to a reduced occurrence of symptomatic heart failure and subclinical cardiac damage. The liposomal formulation of anthracyclines modifies their pharmacokinetics and tissue distribution while maintaining antitumor effectiveness, and it has also been shown to lessen cardiotoxicity.

Dexrazoxane is the sole cardioprotective agent approved by the US Food and Drug Administration for preventing anthracycline-induced cardiotoxicity. Furthermore, antihypertensive medications such as β-blockers, ACE inhibitors, ARBs, and statins have been evaluated in randomized controlled trials for their potential in the primary prevention of this type of cardiotoxicity. Patients who experience symptomatic heart failure should receive prompt and aggressive treatment under the guidance of a cardiologist, following the current standard care protocols.

Oncologists must be aware of the cardiotoxic potential of certain cancer therapies and should promptly assess and monitor any cardiac events that arise during treatment. Long-term follow-up is essential for some therapies, as cardiovascular toxicities, including left ventricular dysfunction or CHF, may manifest more than 10 years after the initial treatment.

Gonadal toxicity

Improvement and better delivery of anticancer agents have greatly improved treatment efficacy and survival rates among cancer patients, but gonadal toxicity remains a big problem. Gonadotoxic antineoplastic agents, including alkylating agents used during reproductive age, can lead to premature ovarian insufficiency and loss of fertility.

In past years, the demand for fertility preservation has risen among female cancer patients. At present, embryo and oocyte cryopreservation are the established options for fertility preservation in women. In women with cancer, fertility loss can be prevented through individualized preservation options, including established techniques, while taking into consideration the patient's age, marital status, chemotherapy regimen, and the possibility of treatment delay.

Effective multidisciplinary oncofertility discussions that involve a highly skilled and experienced oncofertility team, which consists of medical oncologists, gynecologists, reproductive biologists, oncologic surgeons, patient care coordinators, psychologists, and research scientists, are necessary to provide cancer patients with high-quality care (Cho et al., 2020).

Fatigue

It is one of the most common symptoms experienced by patients receiving treatment for cancer. It is characterized by a diminution of energy and an excessive need to rest. Cancer-related fatigue is experienced and reported differently by each patient (Wang et al., 2014).

The underlying mechanisms of CRF are believed to involve several physiological, biochemical, and behavioral systems. There is growing evidence that various factors contribute to changes in corticotropin-releasing factor (CRF). These include:

- Gene polymorphisms: Variations in genes that can affect biological functions.
- HPA axis disruption: Imbalances in the hypothalamic–pituitary–adrenal axis, which regulates the stress response.
- Altered circadian rhythms: Changes in the body's natural biological clock that affect hormone release.
- Diminished heart rate variability: Reduced fluctuations in heart rate, which indicate stress or health issues.
- Immune dysfunction: Impaired immune system responses that can lead to health problems.
- Serotonin dysregulation: Imbalances in serotonin, a neurotransmitter that affects mood and behavior.
- Mitochondrial dysfunction: Impairments in the energy-producing structures within cells.
- 5-Hydroxytryptophan dysregulation: Issues with the precursor of serotonin, impacting mood and other functions.
- Proinflammatory cytokine activity: Increased levels of signaling molecules that promote inflammation.

Together, these mechanisms can significantly influence the regulation of CRF and overall health (Lacourt and Heijnen, 2017; Wang et al., 2017).

Fatigue often results from multiple underlying causes in each patient, making it crucial to tailor treatment plans to individual needs. Maintaining open communication among the patient, their family, and the healthcare team can help explore the impact of fatigue on daily life. General supportive care for patients experiencing fatigue includes promoting a balanced diet with sufficient fluids, calories, protein, carbohydrates, fats, vitamins, and minerals. It's also important to find a balance between rest and physical activity, along with engaging in restorative activities like spending time in nature or enjoying uplifting distractions such as music (Mustian et al., 2017; Pearson et al., 2018).

Physical activity and exercise play a beneficial role, as proven in various metaanalyses.

Hair loss/hair changes

Loss of hair can cause trauma to both male and female cancer patients. Chemotherapy agents that are likely to cause complete alopecia include cyclophosphamide, ifosfamide, doxorubicin, daunorubicin, idarubicin, dactinomycin, paclitaxel, docetaxel, ixabepilone, eribulin, etoposide, and irinotecan. A few other cancer treatments besides cytotoxic agents that can cause hair loss are decitabine, anticancer monoclonal antibodies, bortezomib, everolimus, sunitinib, imatinib, axitinib, dasatinib, crizotinib, erlotinib, pazopanib, trametinib, palbociclib, nilotinib, cabozantinib, etc. (Balagula et al., 2010; Dorr, 1998).

Several compounds, e.g., topical minoxidil, epidermal growth factor, fibroblast growth factors, N-acetylcysteine, interleukin (IL) 1 injection, calcitriol, parathyroid hormone, α-tocopherol, etc., have been studied with mixed results in the prevention and treatment of hair loss.

Cranial irradiation can lead to hair loss. Palliative or prophylactic whole-brain radiation or partial-brain radiotherapy for primary brain tumors causes temporary hair loss, which starts approximately 2–3 weeks after initiating radiation and resolves 2–3 months after the cessation of radiotherapy.

Neurocognitive effects

Cognitive dysfunction is common in many patients with brain tumors, often appearing at the time of diagnosis and typically worsening over time. Cognition plays a crucial role in nearly all daily activities, and even minor changes in cognitive abilities can greatly affect quality of life and independence. A formal neuropsychological assessment offers quantitative and objective measurements of subtle but clinically important changes in cognitive function (Litofsky et al., 2004).

Nonbrain surgery can lead to postoperative cognitive impairments, particularly in older patients. Memory deficits may persist for weeks or even months. Factors like the length of anesthesia and complications after surgery also contribute to these issues. In contrast, surgery for brain tumors can cause or worsen specific cognitive deficits, especially in the short term (Habets et al., 2014).

Radiation therapy is crucial in treating primary and metastatic CNS tumors. Prospective studies have shown that modern focal treatments are generally safe, but risks still exist, particularly with less targeted approaches like whole-brain radiation therapy (Glosser et al., 1997).

Research indicates that patients receiving high-dose chemotherapy experience greater cognitive decline than those undergoing conventional chemotherapy. Furthermore, there appears to be a linear decline in cognitive function after each chemotherapy cycle, suggesting a dose–response relationship (van Dam et al., 1998).

Long-term neurocognitive changes may stem from epigenetic biological alterations in circulating peripheral proinflammatory cytokines, such as IL-6, IL-8, IL-10, and tumor necrosis factor-alpha. These cytokines are released in response to chemotherapy and can trigger a cascade that disrupts the CNS. There is evidence linking cytokine levels to brain volume and memory performance (Wang et al., 2015).

There are few proven therapies for cognitive dysfunction. At the outset, when the diagnosis is established, reversible causes like medications, endocrine disturbances, depression, etc., should be evaluated. Donepezil has shown benefit in a few patients. Brain injury rehabilitation is one of the treatment options (Gehring et al., 2009).

Other nonpharmacological approaches, like exercise programs and mindfulness-based stress reduction, are receiving growing interest as potentially effective methods for addressing cognitive effects related to cancer and its treatment (Johns et al., 2016).

Second cancers

Advancements in contemporary cancer treatments have led to improved survival rates specific to cancer. However, these advancements carry long-term risks, particularly for children, adolescents, and young adults. Subsequent cancers in survivors account for 15%–20% of all cancer cases. The likelihood of developing a second cancer is notably elevated compared to the general population, with a risk that is approximately three times higher.

Certain chemotherapeutic drugs are recognized for their potential to raise the risk of leukemia, typically manifesting shortly after chemotherapy. Notable examples include alkylating agents such as cyclophosphamide and ifosfamide, which have a median onset time for leukemia of 4–6 years posttreatment. The use of epipodophyllotoxins, whether alone or in conjunction with alkylating agents, further heightens the risk of therapy-related leukemia. In addition, topoisomerase inhibitors are linked to AML, with a latency period ranging from 1 to 3 years. Another significant example of therapy-related cancer risk is the use of tamoxifen, a treatment for hormone receptor-positive breast cancer in premenopausal women, which has been associated with an increased risk of endometrial cancer (OR = 1.52; 95% CI = 1.07–2.17).

Predicting who may develop a second cancer is not feasible; however, it is crucial to recognize the potential risks for future health issues and secondary cancers in cancer survivors. Regular follow-up and screening tests play a vital role in ongoing monitoring. A comprehensive cancer survivorship care plan should emphasize the necessity of long-term follow-up and screening to address posttreatment symptoms, recurrence, and the possibility of second cancers (Demoor-Goldschmidt and de Vathaire, 2019).

Cancer survivorship

The diverse needs of survivors require a range of responses from clinicians to ensure a high quality of life after treatment. The essential components of survivorship care were established by the **Committee on Cancer Survivorship from the Institute of Medicine.**

This landmark report outlined specific objectives to promote optimal outcomes for cancer survivors, including:

1. Increase awareness of the medical and psychosocial challenges faced by cancer survivors.
2. Define quality healthcare for cancer survivors and strategies to achieve it.
3. Enhance quality of life by implementing policies that ensure access to psychosocial services, fair employment practices, and health insurance for cancer survivors.

Upon discharge from cancer treatment, each patient should receive a comprehensive record that includes:

1. Contact information: Details for each treating institution and key healthcare providers.
2. Treatment timeline: Dates marking the beginning and completion of treatment.
3. Diagnostic information: Summary of diagnostic tests and their results.
4. Tumor characteristics: Specific features of the tumor.
5. Treatment details: Information on surgeries or therapies received, including agents used, regimens, dosages, the number and titles of any clinical trials participated in, treatment responses, and any toxicities experienced.
6. Support services: Overview of psychosocial and nutritional services provided.
7. Care coordination: Identification of the individual responsible for coordinating ongoing care.

In addition, both the patient and their primary healthcare provider should receive a written follow-up care plan containing:

1. Recovery outlook: Expected course of recovery posttreatment.
2. Follow-up schedule: Recommendations for periodic testing and examinations, including who will conduct them and their schedule.
3. Long-term effects: Information on potential late and long-term effects of treatment and associated symptoms.
4. Recurrence signs: Signs to watch for that may indicate recurrence or the development of secondary tumors.
5. Impact on daily life: A discussion of how cancer may affect personal relationships, work, and mental health.
6. Financial and employment considerations: Information on potential insurance, employment, and financial implications of cancer.
7. Healthy behavior recommendations: Suggestions for healthy lifestyle changes to share with first-degree relatives to reduce their cancer risk.
8. Genetic counseling information: Details on genetic counseling and testing options for high-risk individuals.
9. Chemoprevention strategies: Information on effective chemoprevention methods for secondary prevention.

10. Referrals to specialists: Recommendations for specific follow-up care providers.
11. Resource listings: A compilation of cancer-related print and online resources, as well as support organizations (Anon, 2005).

References

Andreyev, J., Ross, P., Donnellan, C., et al., 2014. Guidance on the management of diarrhoea during cancer chemotherapy. Lancet Oncol. 15 (10), e447–60.

Anon, 2005. From Cancer Patient to Cancer Survivor: Lost in Transition [Internet]. National Academies Press, Washington, D.C. Available from: http://www.nap.edu/catalog/11468. (cited 2025 Mar 8).

Balagula, Y., Lacouture, M.E., Cotliar, J.A., 2010. Dermatologic toxicities of targeted anticancer therapies. J. Support. Oncol. 8 (4), 149–161.

Caillot, D., Couaillier, J.-F., Bernard, A., et al., 2001. Increasing volume and changing characteristics of invasive pulmonary aspergillosis on sequential thoracic computed tomography scans in patients with neutropenia. J. Clin. Oncol. 19 (1), 253–259.

Cho, H.-W., Lee, S., Min, K.-J., et al., 2020. Advances in the treatment and prevention of chemotherapy-induced ovarian toxicity. Int. J. Mol. Sci. 21 (20), 7792.

Citrin, D.E., Shankavaram, U., Horton, J.A., et al., 2013. Role of type II pneumocyte senescence in radiation-induced lung fibrosis. J. Natl. Cancer Inst. 105 (19), 1474–1484.

Cloutier, R.L., 2009. Neutropenic enterocolitis. Emerg. Med. Clin. North Am. 27 (3), 415–422.

Cooley, L., Dendle, C., Wolf, J., et al., 2014. Consensus guidelines for diagnosis, prophylaxis and management of neumocystis jirovecii pneumonia in patients with haematological and solid malignancies, 2014. Intern. Med. J. 44 (12b), 1350–1363.

Demoor-Goldschmidt, C., de Vathaire, F., 2019. Review of risk factors of secondary cancers among cancer survivors. Br. J. Radiol. 92 (1093), 20180390.

Dorr, V.J., 1998. A practitioner's guide to cancer-related alopecia. Semin. Oncol. 25 (5), 562–570.

Flowers, C.R., Seidenfeld, J., Bow, E.J., et al., 2013. Antimicrobial prophylaxis and outpatient management of fever and neutropenia in adults treated for malignancy: American Society of Clinical Oncology Clinical Practice Guideline. J. Clin. Oncol. 31 (6), 794–810.

Gehring, K., Sitskoorn, M.M., Gundy, C.M., et al., 2009. Cognitive rehabilitation in patients with gliomas: a randomized, controlled trial. J. Clin. Oncol. 27 (22), 3712–3722.

Glosser, G., McManus, P., Munzenrider, J., et al., 1997. Neuropsychological function in adults after high dose fractionated radiation therapy of skull base tumors. Int. J. Radiat. Oncol. Biol. Phys. 38 (2), 231–239.

Graves, P.R., Siddiqui, F., Anscher, M.S., Movsas, B., 2010. Radiation pulmonary toxicity: from mechanisms to management. Semin. Radiat. Oncol. 20 (3), 201–207.

Gupta, K., Walton, R., Kataria, S.P., 2021. Chemotherapy-induced nausea and vomiting: pathogenesis, recommendations, and new trends. Cancer Treat. Res. Commun. 26, 100278.

Habets, E.J.J., Kloet, A., Walchenbach, R., Vecht, C.J., Klein, M., Taphoorn, M.J.B., 2014. Tumour and surgery effects on cognitive functioning in high-grade glioma patients. Acta Neurochir. (Wien) 156 (8), 1451–1459.

Haim, N., Epelbaum, R., Ben-Shahar, M., Yarnitsky, D., Simri, W., Robinson, E., 1994. Full dose vincristine (without 2-mg dose limit) in the treatment of lymphomas. Cancer 73 (10), 2515–2519.

Inomata, S., Takahashi, H., Nagata, M., et al., 2004. Acute lung injury as an adverse event of gefitinib. Anticancer Drugs 15 (5), 461–467.

Johns, S.A., Brown, L.F., Beck-Coon, K., et al., 2016. Randomized controlled pilot trial of mindfulness-based stress reduction compared to psychoeducational support for persistently fatigued breast and colorectal cancer survivors. Support. Care Cancer 24 (10), 4085–4096.

Lacourt, T.E., Heijnen, C.J., 2017. Mechanisms of neurotoxic symptoms as a result of breast cancer and its treatment: considerations on the contribution of stress, inflammation, and cellular bioenergetics. Curr. Breast Cancer Rep. 9 (2), 70–81.

Ladhani, S.N., Fernandes, S., Garg, M., et al., 2024. Prevention and treatment of infection in patients with an absent or hypofunctional spleen: a British Society for Haematology guideline. Br. J. Haematol. 204 (5), 1672–1686.

Lin, J.-T., Yeh, K.-T., Fang, H.-Y., Chang, C.-S., 2006. Fulminant, but reversible interstitial pneumonitis associated with imatinib mesylate. Leuk. Lymphoma 47 (8), 1693–1695.

Litofsky, N.S., Farace, E., Anderson, F.J., et al., 2004. Depression in patients with high-grade glioma: results of the glioma outcomes project. Neurosurgery 54 (2), 358.

Maertens, J.A., Raad, I.I., KA, M., et al., 2016. Isavuconazole versus voriconazole for primary treatment of invasive mould disease caused by Aspergillus and other filamentous fungi (SECURE): a phase 3, randomised-controlled, non-inferiority trial. Lancet 387 (10020), 760–769.

Mermel, L.A., Allon, M., Bouza, E., et al., 2009. Clinical practice guidelines for the diagnosis and management of intravascular catheter-related infection: 2009 Update by the Infectious Diseases Society of America. Clin. Infect. Dis. 49 (1), 1–45.

Mustian, K.M., Alfano, C.M., Heckler, C., et al., 2017. Comparison of pharmaceutical, psychological, and exercise treatments for cancer-related fatigue a meta-analysis. JAMA Oncol. 3 (7), 961–968.

Nichols, W.G., Price, T.H., Gooley, T., Corey, L., Boeckh, M., 2003. Transfusion-transmitted cytomegalovirus infection after receipt of leukoreduced blood products. Blood 101 (10), 4195–4200.

Pearson, E.J.M., Morris, M.E., Di Stefano, M., McKinstry, C.E., 2018. Interventions for cancer-related fatigue: a scoping review. Eur. J. Cancer Care (Engl.) 27 (1), e12516.

Punekar, S.R., 2020. Chapter 4—cancer treatment-related thrombocytopenia [Internet]. In: Velcheti, V., Punekar, S. (Eds.), Handbook of Cancer Treatment-Related Symptoms and Toxicities. Elsevier, Philadelphia, pp. 17–20. Available from: https://www.sciencedirect.com/science/article/pii/B9780323672412000045. (cited 2025 Mar 8).

Roden, M.M., Zaoutis, T.E., Buchanan, W.L., et al., 2005. Epidemiology and outcome of zygomycosis: a review of 929 reported cases. Clin. Infect. Dis. 41 (5), 634–653.

Swaika, A., Paulus, A., Miller, K.C., et al., 2012. Acyclovir prophylaxis against varicella zoster virus reactivation in multiple myeloma patients treated with bortezomib-based therapies: a retrospective analysis of 100 patients. J. Support. Oncol. 10 (4), 155–159.

Tarricone, R., Abu Koush, D., Nyanzi-Wakholi, B., Medina-Lara, A., 2016. A systematic literature review of the economic implications of chemotherapy-induced diarrhea and its impact on quality of life. Crit. Rev. Oncol. Hematol. 99, 37–48.

Theis, V.S., Sripadam, R., Ramani, V., Lal, S., 2010. Chronic radiation enteritis. Clin. Oncol. (R. Coll. Radiol.) 22 (1), 70–83.

van Dam, F.S.A.M., Boogerd, W., Schagen, S.B., et al., 1998. Impairment of cognitive function in women receiving adjuvant treatment for high-risk breast cancer: high-dose versus standard-dose chemotherapy. J. Natl. Cancer Inst. 90 (3), 210–218.

Wang, T., Yin, J., Miller, A.H., Xiao, C., 2017. A systematic review of the association between fatigue and genetic polymorphisms. Brain Behav. Immun. 62, 230–244.

Wang, X.-M., Walitt, B., Saligan, L., Tiwari, A.F., Cheung, C.W., Zhang, Z.-J., 2015. Chemobrain: a critical review and causal hypothesis of link between cytokines and epigenetic reprogramming associated with chemotherapy. Cytokine 72 (1), 86–96.

Wang, X.S., Zhao, F., Fisch, M.J., et al., 2014. Prevalence and characteristics of moderate to severe fatigue: a multi-center study in cancer patients and survivors. Cancer 120 (3), 425–432.

CHAPTER

9

Preclinical studies: Translational research in drug development

Sankar Jyoti Bora[a], *Santeena Chensula*[b], *Anusuya Bhattacharyya*[d], *Phulen Sarma*[a], *and Parag Deb Roy*[c]

[a]Department of Pharmacology, AIIMS, Guwahati, India
[b]Department of Clinical Pharmacology, AIIMS, Bhopal, India
[c]Integrated Disease Surveillance Programme, Office of the Joint Director of Health Services, Guwahati, India
[d]Department of Ophthalmology, AIIMS, Guwahati, India

The chapter highlights the integration of multiomics technologies and biomarker-driven patient stratification in optimizing target identification, drug repurposing, and precision treatment strategies. While existing models have limitations, such as the inability of patient-derived xenograft (PDX) to assess immunotherapies, emerging solutions such as humanized mouse models and immune-competent organ-on-a-chip platforms are improving translational success. By fostering robust academia–industry collaborations and implementing adaptive clinical trial designs, translational research can accelerate the development of precision therapies, reduce late-stage failures, and improve patient outcomes. This comprehensive review underscores the transformative power of translational paradigms in overcoming longstanding challenges in oncology, paving the way for more predictive, individualized, and clinically impactful cancer therapeutics.

Introduction

The goal of the intricate and demanding process of developing cancer drugs is to find and provide patients with efficient treatments. The worldwide burden of cancer, which continues to be a major cause of morbidity and mortality, as well as the complexity and heterogeneity of the disease, which includes genetic alterations, altered signaling pathways, and immune evasion mechanisms, makes such advancement necessary Rubin and Gilliland, 2012). Preclinical research usually starts the development process. High-throughput assays,

Pharmacological Approaches in Cancer Treatment
https://doi.org/10.1016/B978-0-443-38307-6.00008-9

computational modeling, and in vitro and in vivo investigations are used to identify possible drug candidates and evaluate their safety and efficacy (Szymański et al., 2012; Wu et al., 2020). To improve their pharmacokinetics, bioavailability, and target specificity, lead compounds are optimized. After that, promising drugs go through Phase I (safety evaluation), Phase II (efficacy assessment), and Phase III (comparative studies in larger populations) of clinical trials. If the medication shows substantial benefits with manageable risks, regulatory approval comes next. Postmarketing surveillance (Phase IV) ensures long-term safety and effectiveness (Biala et al., 2023). Few medications make it to clinical trials, receive approval, and hit the market despite the time, labor, and financial expenditure (Sun et al., 2022). The initial research that assesses a lead molecule's safety and effectiveness is known as a non-clinical/preclinical study. Since preclinical research findings will inform subsequent clinical methods, the data produced by preclinical research are critical to the licensure and success of a lead drug. Therefore it is vital to choose preclinical models that may yield results that can be applied to clinical studies with a higher chance of being authorized as a novel medicine. Strict optimization in preclinical models with an eye on the product's eventual usage by patients can lower failure rates and finally accomplish the intended goal (Wittenburg and Gustafson, 2011).

To promote human health, translational research—also referred to as translational science or translation—connects fundamental research discoveries with practical applications and patient care. Because it aims to bring laboratory discoveries to real-world patient applications, it is frequently referred to as bench-to-bed research. Translational research consists of two stages, which disseminate the initial laboratory results to the general public. Knowledge from laboratory research is transferred to clinical research in the first phase, or T1, and clinical research findings are transferred to medical practice and communities to improve health in the second phase, or T2 (Rubio et al., 2010). Translational research is a paradigm that ensures the majority of preclinical ideas reach clinical trials and have the best chance of succeeding in terms of safety and societal benefits (Choi et al., n.d.). It raises the success rate of cancer drug development by combining preclinical validation, biomarker-based precision medicine, adaptive clinical trials, and AI-driven insights, which eventually result in faster and better treatments for patients. Its primary goal is to increase community acceptance of medication molecules for use (Zang and Lee, 2014; Alum, 2025). Translational drug discovery demands both cooperation between clinical and pharmacological research and the significance of the role of academia in target identification and drug discovery, design, and development. Translational drug discovery requires collaboration between clinical and pharmacological research, which helps with the critical phases of target identification and drug discovery, design, and development (Researchgate, 2025). Translational research can address all of the issues surrounding medication development and discovery, including the previously mentioned failure of clinical tests. In contrast to traditional drug development, translational drug development incorporates cutting-edge scientific approaches to improve clinical relevance and efficiency. Translational research uses patient-derived organoids (PDOs) and multiomics technologies (genomics, proteomics, and metabolomics) to increase specificity, in contrast to conventional methods that depend on in vitro tests and animal models for target validation (Yang and Yu, 2023; Chen et al., 2023). While translational research uses three-dimensional (3D) cell cultures, humanized mouse models, and organ-on-a-chip systems for enhanced human applicability, preclinical models in traditional drug development mostly use standard cell cultures and animal testing. While translational research uses biomarker-driven patient stratification, adaptive clinical trial designs, and real-time monitoring to increase success rates, traditional drug development in clinical trials takes a strict, phase-based approach (Fang and Eglen, 2017; Park et al., 2025; Mosig, 2017). Drug repurposing and precision medicine techniques such as genetic profiling and microbiome analysis to maximize therapeutic outcomes for individual patients are integral parts of translational research (De Maria Marchiano et al., 2021; Petrosino, 2018). Traditional drug development relies on generalized pharmacokinetic and pharmacodynamic models. By connecting cutting-edge preclinical models, translational research can support the development of cancer drugs and increase clinical success. This method can speed up the development of tailored therapeutics, which eventually results in more individualized and efficient cancer treatments. This discussion will concentrate on 3D cancer models and humanized animal models, which have greatly improved translational cancer research by offering more physiologically realistic systems for examining tumor biology and treatment outcomes. While 3D cancer models, such as organoids and tumor-on-a-chip systems, more accurately mimic the tumor microenvironment (TME) and medication responses, humanized animal models allow the assessment of human immune–tumor interactions, especially in immunotherapy research. These models provide better insights into the course of cancer and treatment options by addressing significant shortcomings of conventional preclinical methods.

Humanized animal models

The recapitulation of the complex TME is required, as conventional animal models and two-dimensional (2D) in vitro examinations are failing to achieve this. The concept of more humanized animal models comes into play. Animal models of clinical relevance can be used to replicate the treatment of cancer patients in real life, including issues such as resistance, recurrence, and metastasis. Therefore to improve translational success to the clinic, in vivo/ in vitro to human predictability must be established early in the drug development process. Furthermore, both humans and animals are complex systems that are inherently unpredictable because they are more than the sum of their parts. Choosing predictive and verified animal models that are suited to particular clinical problems is essential. Thus preclinical models need to be upgraded in line with advancements in biomedical research (Leenaars et al., 2019; Wege, 2018).

Patient-derived tumor xenograft models

A piece of living tissue that can be transplanted from one living system to another living system is called a graft. Xenografts are living tissues that are transplanted from one species to another, unlike allografts, which are done within the same species. Xenograft mouse models are an important tool in preclinical cancer research. They involve the transfer of human cancer cell lines or the transplantation of fresh PDX into immunocompromised animals (Okada et al., 2018). In contrast to conventional models that employ well-established cancer cell lines, PDX models preserve the richness and diversity of the original tumor by retaining its genetic and molecular features. These models rely on critical components, including the patient-derived tumor tissue, which maintains the unique properties of the original cancer, and specialized immunodeficient animals, such as NOD scid gamma (NSG) or nude mice, which lack the ability to reject human cells. Tumor behavior and response to treatment are also significantly influenced by the TME, which is made up of blood vessels, extracellular matrix (ECM) components, and supportive cells like fibroblasts. To create a PDX model, tumor fragments are carefully implanted either beneath the skin or into the primary organ of the malignancy, allowing them to grow in a manner similar to their normal course. After the tumor has taken root, scientists can grow it by transplanting tissue into more mice for analysis, as shown in Fig. 9.1 Rubin and Gilliland, 2012; Pompili et al., 2016).

These models are frequently used to test cancer medications, investigate resistance mechanisms, and identify possible biomarkers that could indicate how well a treatment will work. PDX models, which closely resemble human tumors, aid in the advancement of personalized medicine by guiding researchers toward more efficient treatments that are customized for each patient. Because they closely resemble human tumors, PDX models are important in cancer research for drug resistance studies, treatment testing, and identifying biomarkers that predict a patient's response to treatment. These models assist researchers in assessing immunotherapies, targeted treatments, and chemotherapy prior to clinical trials by maintaining the genetic and molecular characteristics of the original tumor. In addition, they shed light on the reasons behind some tumors' resistance to treatment,

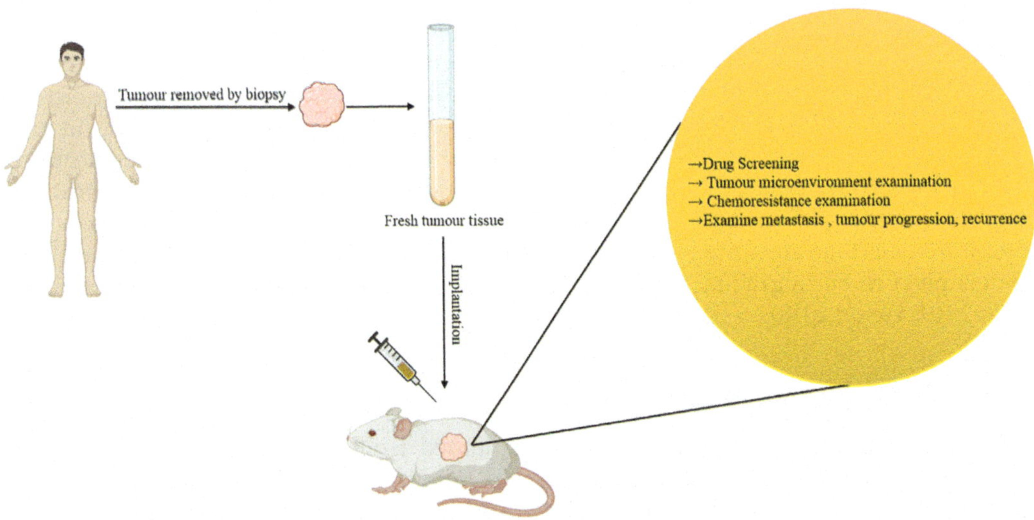

FIG. 9.1 Patient-derived tumor xenograft models. *From Rubin, E.H., Gilliland, D.G., (2012). Drug development and clinical trials--the path to an approved cancer drug. Nat. Rev. Clin. Oncol. 9 (4), 215-22. doi:10.1038/nrclinonc.2012.22. PMID: 22371130.*

enabling researchers to create more effective countermeasures (Liu et al., 2023; Yoshida, 2020). Because PDX models can inform therapy choices based on a patient's tumor features, they are highly valuable in personalized medicine. They also assist researchers in investigating novel immunotherapy alternatives when used with humanized mouse models. All things considered, these models are essential to translating research findings from laboratories into effective cancer therapies, which ultimately enhances patient care and outcomes. PDX models possess genetic and pathological characteristics similar to those of primary tumors. Moreover, PDX models have the ability to mimic the TME of the original tumor (Xie and Lin, 2020).

Human hepatocellular carcinoma (HCC) samples were used to create a PDX model, which effectively screened the cancer's progression while maintaining the cancer's physiology in vivo. An overall success rate of about 20% was achieved when 54 human HCC needle biopsies from patients were subcutaneously transplanted into immunodeficient, nonobese, diabetic, and severe combined immunodeficiency gamma-c mice. Over six generations of retransplantation, HCC PDX tumors maintained the histological, transcriptomic, and genomic features of the initial HCC biopsies. Although the transplanted tumors retained the features of the original tumors despite serial passaging and represented the full range of the genetic landscape of HCCs, the use of immunodeficient mice makes them unsuitable for immunotherapy testing (Blumer et al., 2019). Standard chemotherapy is required to treat triple-negative breast cancer (TNBC), a very aggressive disease that is insensitive to hormone suppression. Neoadjuvant chemotherapy (NACT) is frequently used in this situation, where the medication is given prior to the tumor being surgically removed. Nevertheless, many patients become resistant to this method, which makes total tumor removal difficult (Echeverria et al., 2019a). Taking into account chemoresistance, the factors underlying chemoresistance in TNBC were also examined by researchers using a PDX model. They found that remaining tumors retained their original subclonal architecture following treatment with doxorubicin and cyclophosphamide using PDX models and serial biopsies from TNBC patients receiving NACT. In contrast to untreated tumors, these remaining tumors displayed unique transcriptomic, proteomic, and histological characteristics. The remnant tumors returned to a cyclophosphamide-sensitive condition after the cyclophosphamide treatment ended, suggesting that the drug-tolerant state was reversible and not the result of clonal selection. Biopsies of the patients also revealed this reversible chemoresistance (Echeverria et al., 2019b). Dysregulated ribosomal protein expression and translation lead to breast cancer metastases. Researchers found that metastatic cancer cells have altered ribosome biogenesis and enhanced protein translation, which speeds up tumor growth (Elhamamsy et al., 2022). Using PDX models, they demonstrated that increased levels of ribosomal proteins, particularly RPL15, induce greater translation of prometastatic components. Ribosome profiling indicates that mRNAs encoding proteins involved in cell motility, invasion, and survival are selectively translated by metastatic cells. Hyperactive ribosome activity, which mechanistically raises the expression of epithelial–mesenchymal transition (EMT) regulators, including SNAIL and ZEB1, promotes the possibility of metastasis. These findings suggest that targeting translational machinery is one potential therapeutic strategy to stop breast cancer metastasis (Isaacs and Coffey, 1981; Zhao et al., 2021; Ebright et al., 2020).

Genetically engineered mouse models

A useful substitute for xenograft models in the study of cancer biology and the assessment of treatment approaches are genetically engineered mouse models (GEMMs). To build these models, specific genetic mutations associated with human tumors are introduced. In contrast to xenografts, GEMMs mimic the course of natural tumors by developing tumors spontaneously in an immune-competent environment. They are very useful for examining tumor growth and treatment responses because they exhibit genetic diversity and can naturally progress to metastatic stages. GEMMs are among the most advanced models for studying cancer. They are a good tool for investigating intrinsic and acquired drug resistance, cancer evasion, and cancer metastasis. Conditional models employing the Cre–LoxP system for regulated gene activation, transgenic approaches (introducing oncogenes), knockouts (removing tumor suppressor genes), and knock-ins (introducing specific mutations) are used to create GEMMs (Lampreht Tratar et al., 2018; Kersten et al., 2017).

The transforming growth factor beta (TGF-β) pathway, which controls cell proliferation, differentiation, and apoptosis (programmed cell death), depends critically on SMAD4 (Zhao et al., 2018). A GEMM was used to analyze the role of SMAD4 a central mediator of the TGF-β/BMP signaling pathway in the development of prostate cancer (PCa). The GEMM elucidated how the loss of SMAD4 may influence tumor progression and metastasis. By conditionally eliminating the tumor suppressor gene Pten in the prostate epithelium, the researchers created a model that resulted in indolent PCas. They also introduced a Smad4 deletion in these mice to investigate the effects of SMAD4. PCa in the resultant Pten/Smad4 double-knockout animals became extremely aggressive, invasive, and metastasized. Similar to the metastatic pattern observed in human PCa, the study discovered that the absence of SMAD4 markedly boosted tumor invasion and metastasis, with the disease spreading locally to the seminal vesicles, bladder wall, and urethra and distantly to the lymph nodes, lungs, liver, and bones. According to mechanistic research, these cancers' increased EMT was caused by the overexpression of SPP1 (osteopontin), which is known to encourage metastasis. Furthermore, the tumors showed elevated cyclin D1 expression, which resulted in unchecked growth, and apoptotic resistance accelerated the course of the illness (Ding et al., 2011). The findings highlight the significance of SMAD4 as a tumor suppressor and show that its absence accelerates cancer metastases, treatment resistance, and cancer dissemination. This GEMM provides valuable insights into tumor invasion, metastasis, and chemoresistance, which could lead to the development of new antimetastatic therapies and improve methods for treating PCa. This model-informed progression investigation, in conjunction with genetic, functional, and translational studies, identified SMAD4 as a critical regulator of PCa development in mice and humans.

One of the most aggressive and deadly tumors, pancreatic ductal adenocarcinoma (PDAC), has been extensively studied using the (Kirsten Rat Sarcoma Viral Oncogene Homolog) KRAS-driven GEMM (Hingorani et al., 2005). The Cre–loxP method, which permits conditional activation of oncogenic $KRAS^{G12D}$ only in pancreatic cells, was used to produce this model. Pancreatic intraepithelial neoplasia, which is a precursor lesion of PDAC, is formed when pancreas-specific recombination is ensured by the use of Pdx1–Cre or Ptf1a–Cre. These early lesions, however, develop slowly, indicating that more genetic changes are necessary for complete malignant transformation (Distler et al., 2014). In addition to $KRAS^{G12D}$, researchers included mutations in tumor suppressor genes such as Trp53 (p53) and Ink4a/Arf to increase the metastatic potential and more closely resemble human PDAC (Rivlin et al., 2011). While $KRAS^{G12D}$ + Ink4a/Arf deletion also caused fast-progressive PDAC with extensive dissemination, $KRAS^{G12D}$ + $Trp53^{R172H}$ mutation produced extremely aggressive PDAC with frequent metastases to the liver, lungs, peritoneum, and lymph nodes. These models' metastatic behavior, in which cancers invade nearby structures before moving on to other organs, is quite similar to that of human PDAC. These mice's disease course is similar to that of human pancreatic cancer; late-stage PDAC is characterized by highly invasive tumors that spread to lymph nodes, the liver, the lungs, and the peritoneum. The research of PDAC development, metastasis, and treatment resistance has benefited greatly from these models. The $KRAS^{G12D}$ + Trp53-mutant mouse has shown resistance to gemcitabine, the common treatment for PDAC, in drug testing, mirroring comparable therapeutic difficulties encountered in human patients (Natu and Nagaraju, 2023). Therapeutic approaches that target fibrosis and immune evasion have also been developed as a result of the use of these models to investigate interactions between cancer cells and the TME. All things considered, the KRAS-driven mouse model has yielded important new information about the mechanisms underlying PDAC genesis, development, metastasis, and resistance to treatment. Researchers have produced a highly aggressive preclinical model that closely resembles human pancreatic cancer by adding extra mutations like Trp53 and Ink4a/Arf deletions. This makes the model an invaluable tool for drug development, metastasis studies, and evaluating new therapeutic approaches.

Using a specifically modified mouse model, researchers investigated the effects of Neu (HER2/ErbB2) gene activation on the growth and metastasis of breast cancer. To turn Neu expression on and off in the mammary glands, researchers created transgenic mice (Moody et al., 2002). Aggressive breast cancers that eventually migrated to the lungs were caused by the gene when it was activated. They therefore inhibited Neu expression and examined its impact to ascertain whether ongoing Neu signaling was necessary for tumor growth (Lower et al., 2009). The results demonstrated that the tumors' growth required continuous oncogene activity, as the tumors completely shrank when Neu was turned off. Curiously, metastatic cancers in the lungs also receded, albeit more slowly than the original breast tumors. This implied that Neu was still necessary even for the disseminated tumors and that their shrinkage was associated with a higher rate of apoptosis, or the death of cancer cells. This study emphasizes the idea of oncogene addiction, which describes how some tumors significantly depend on particular genetic cues to survive and proliferate. These results have significant therapeutic ramifications, highlighting the value of HER2-targeted treatments such as lapatinib and trastuzumab (Herceptin), which inhibit HER2 signaling to stop cancer metastases and recurrence in patients with HER2-positive breast cancer (Swain et al., 2023; Swain et al., 2023).

A study examined the effectiveness of a GEMM as a preclinical tool for researching and testing potential treatments for HER2-positive breast cancer. The HER2/neu oncogene was overexpressed in mammary epithelial cells in a transgenic mouse model, leading to spontaneous tumor growth that closely mimicked the progression of HER2-positive breast cancer in humans (Creedon et al., 2016). The tumors in these mice, which closely resembled human HER2-positive breast cancer due to the overexpression of the HER2/neu oncogene, enabled researchers to monitor the evolution of tumor response and resistance to trastuzumab and lapatinib over time (Yuan et al., 2022). When malignancies initially regressed but later recurred, researchers analyzed the resistant tumors and identified significant molecular mechanisms, including PI3K/AKT/mTOR activation, PTEN loss, EMT, and cyclin D1 overexpression. These findings facilitated the identification of several survival strategies employed by tumors to evade HER2 suppression. Combination therapy was also evaluated using the model, demonstrating that Cyclin-Dependent Kinase (CDK) and Phosphoinositide 3-kinase (PI3K)/mechanistic target of rapamycin (mTOR) inhibitors could improve treatment responsiveness and delay resistance. This study highlighted the critical role of HER2-targeted therapy in tumor maintenance and provided valuable insights for developing innovative strategies to overcome drug resistance in HER2-positive breast cancer (Creedon et al., 2016).

3D cancer models

In recent years, with the increasing research on tissue engineering, in vitro 3D culture models have received more and more attention from researchers. 3D cancer models are in vitro models that can easily mimic the tumor TME. There are proven translational potentials of 3D cancer models such as tumor spheroids, organoids, and immune tumor-on-a-Chip models. Compared to traditional 2D cell cultures, 3D cancer models are complex laboratory systems that more closely resemble tumor composition and characteristics. These models mimic genuine tumor conditions by providing a 3D environment in which cancer cells thrive and interact with surrounding tissues. These models are valuable for preclinical research and tailored treatment because they allow researchers to investigate tumor biology, medication responses, metastasis, and therapy resistance in a manner that is similar to human cancer (Manduca et al., 2023; Guan and Huang, 2022; Sharma et al., 2024).

Tumor spheroids

Tumor spheroids are spherical aggregates of cancer cells that mimic the physiology of solid tumor such as internal structures, cell signaling, ECM, and cellular heterogeneity (Roy et al., 2023). There are basically four types of tumor spheroids; among them, multicellular tumor spheroids (MCTSs) are the most commonly used. Others include tumorospheres, tissue-derived tumorospheres, and organotypic multicellular spheroids. Among all tumor spheroids, MCTSs are more efficient in mimicking the TME as they have heterogeneous, multilayered cell organization (Mitrakas et al., 2023). The outer layer generally consists of proliferating cells, with a middle core layer of hypoxic environment bearing senescent cells and inner most necrotic cells having acidic properties, as shown in Fig. 9.2 (Szymański et al., 2012). Each cell layer of the spheroids provides meaningful insights toward drug action. The hypoxic core of cells within in vivo tumors and cells in 3D cultures has been shown to exhibit resistance toward anticancer drugs (Pinto et al., 2020). The hypoxic monolayer of glioma and astrocytoma cell tumor spheroids exhibited higher resistance to doxorubicin, which is a well-known inducer of apoptosis, by lowering the production of BCL-2. This highlighted the antiapoptotic activity of hypoxic cells present in the TME and provided an

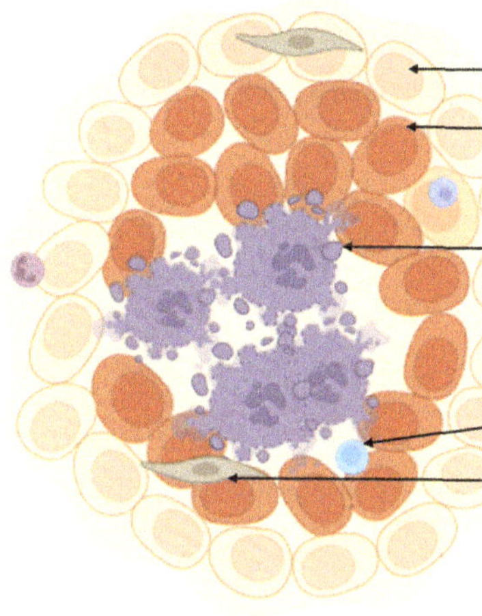

FIG. 9.2 Tumor spheroids. *From Szymański, P., Markowicz, M., Mikiciuk-Olasik, E., 2012. Adaptation of high-throughput screening in drug discovery—toxicological screening tests. Int. J. Mol. Sci. 13 (1), 427–452. doi:10.3390/ijms13010427.*

in vivo-like environment to validate the point. Similar types of doxorubicin resistance were observed in MC75 breast cancer cell tumor spheroids (Al-Akra et al., 2019; Doublier et al., 2012).

Tumor organoids

Organoids are 3D, organ-like structures that are produced in a laboratory using pluripotent stem cells or adult tissues (Yang et al., 2023). These organoids are referred to as PDOs since the tissues or cells are human. Although organoids can also be made from animal tissues, the focus is more on PDOs because translational research and the imitation of human organs have been prioritized. Comparing these organoid models to mouse PDX establishment, one of their main advantages is their high efficiency and comparatively low cost. PDX models are made by directly implanting patient tumor fragments into immunocompromised mice, allowing the tumor to grow in vivo while maintaining its original characteristics (Wensink et al., 2021). Organoids are miniature, 3D, tumor-like structures made from patient-derived tumor cells and cultured in vitro. Compared to spheroids, they are more physiologically relevant. Organoids obtained from patients can be produced using biopsies of tumor tissue. These tumor organoids, also known as tumoroids, have been shown to retain key genetic and behavioral characteristics of their tumor subtype and the tissue of origin, while also maintaining intratumoral heterogeneity (Abbasi-Malati et al., 2024). Organoid specimens can be obtained through piercing or surgical biopsies, after which individual cells are separated by enzymatic and mechanical means and then cultivated. They are cultivated in specific media that support their 3D structure and cellular development, as shown in Fig. 9.3 (Wu et al., 2020). A basal medium enriched with B27 and N-acetylcysteine often

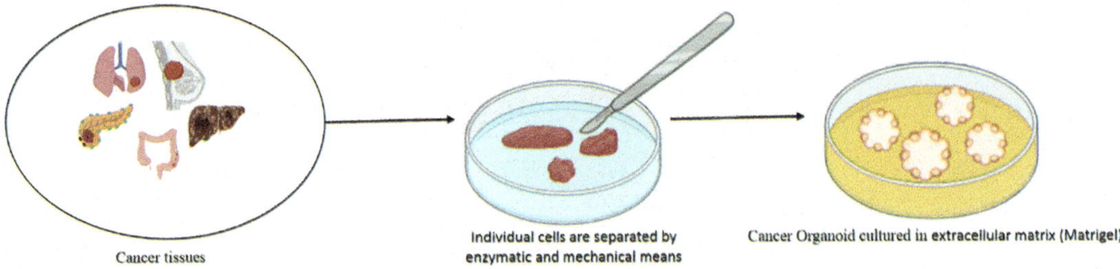

FIG. 9.3 Tumor organoids. *From Wu, F., Zhou, Y., Li, L., Shen, X., Chen, G., Wang, X., Liang, X., Tan, M., Huang, Z., 2020. Computational approaches in preclinical studies on drug discovery and development. Front. Chem. 8, 2296-2646. Available from: http://journal.frontiersin.org/journal/chemistry. doi:10.3389/fchem.2020.00726.*

makes up the culture medium, along with growth factors such as EGF, Noggin, R-spondin, and Wnt ligands. The organoids are inserted using extracellular matrices, such as Matrigel, which mimic the tissue milieu. These medium formulations maintain the viability and functionality of organoids. The nutritional media for organoid cultures are determined by the type of cell and the tissue of origin (Zhao et al., 2022).

A living biobank of breast cancer organoids—laboratory-grown 3D models made from patient tumors—was produced in a study. These organoids are useful for researching breast cancer in a manner that closely resembles the disease in patients since they preserve the genetic and structural variety of actual tumors. Researchers discovered important cancer-related mutations, including TP53, PIK3CA, and GATA3, and variations in gene expression connected to different breast cancer subtypes by examining their DNA, RNA, and proteins. E-cadherin, a protein that aids in cell adhesion, was lost in some organoids, which may indicate a role in the spread of cancer. The organoids showed responses that were comparable to those observed in patients when tested with various medications. For example, organoids that were positive for the estrogen receptor (ER) responded to tamoxifen, a standard treatment for breast cancer, whereas those with PIK3CA mutations were sensitive to PI3K inhibitors. In addition, chemotherapy medications such as paclitaxel were evaluated, shedding light on drug resistance. These organoids present a promising approach for creating customized therapeutics and enhancing the treatment of breast cancer by maintaining the intricacy of actual tumors and their responses to therapy (Sachs et al., 2018; Bruner and Derksen, 2018; Gruvberger-Saal et al., 2007).

Researchers used patient-derived organoids of breast cancer patients to reveal new mechanistic insights into breast cancer progression and drug efficacy. A study explores how fibroblasts, which are the supporting cells of breast tissues, release cytokine signals like IL-1β that help in ER-positive breast cancer growth. Cytokines such as CCL7, IL-6, and IL-8 are constitutively secreted by fibroblasts, and they cause ER^+ breast cancer cells to generate platelet-derived growth factor BB (PDGF-BB). Fibroblasts are then signaled by PDGF-BB to create interleukin-1 beta (IL-1β), which results in an IL-1β-enriched microenvironment that encourages the growth of cancer cells. According to this method, cancer recurrence may be triggered by normal fibroblasts in tumor-adjacent tissue following breast-conserving surgery. Targeting the PDGF-BB and IL-1β pathways in addition to tamoxifen therapy may enhance treatment outcomes for patients with ER^+ breast cancer. The study also shows that tamoxifen is more effective in suppressing tumor cell proliferation when this signaling is suppressed (Chatterjee et al., 2019).

Tumor-on-a-chip model

Tumor-on-a-chip model is a type of microfluidic organ-on-a-chip designed to mimic the TME in vitro, as shown in Fig. 9.4 (Biala et al., 2023). A microfluidic environment platform is used to integrate cells, ECM, normoxia, hypoxia, nutrients, vascularization, and immune systems (Liu et al., 2021). Researchers were able to monitor the development of tumors in real time by introducing normal tissue cells and colon cancer cells into a hydrogel that could be customized. The hydrogel mimicked the ECM. A circulating fluid system that mimics blood flow connects two chambers, one of which represents the gut and the other the liver, in the "metastasis-on-a-chip" system, which is intended to enhance cancer research and drug discovery. The cancer cells proliferated rapidly, lost adhesion markers, changed to a metastatic phenotype, and ultimately moved from the gut to the liver chamber. By enabling drug testing and chemical changes to the hydrogel, the technology aids researchers in understanding how cancers spread to new tissues and react to therapy. This novel model offers a more regulated and realistic setting for researching metastases and creating novel cancer treatments (Skardal et al., 2016). Because they closely resemble the natural ECM, hydrogels are essential in metastasis-on-a-chip systems as they offer a 3D environment that promotes realistic cell behavior. In these systems, hydrogels facilitate the creation of 3D cell cultures, which are crucial for precisely simulating tumor growth and metastasis because they enable cells to grow and interact in all directions. Changing the hydrogel's stiffness can affect the invasion and migratory patterns of tumor cells, providing information on the physical signals that promote metastasis. Hydrogels can be incorporated into microfluidic devices to produce intricate tissue models in a regulated setting. The adaptability of hydrogel formulations

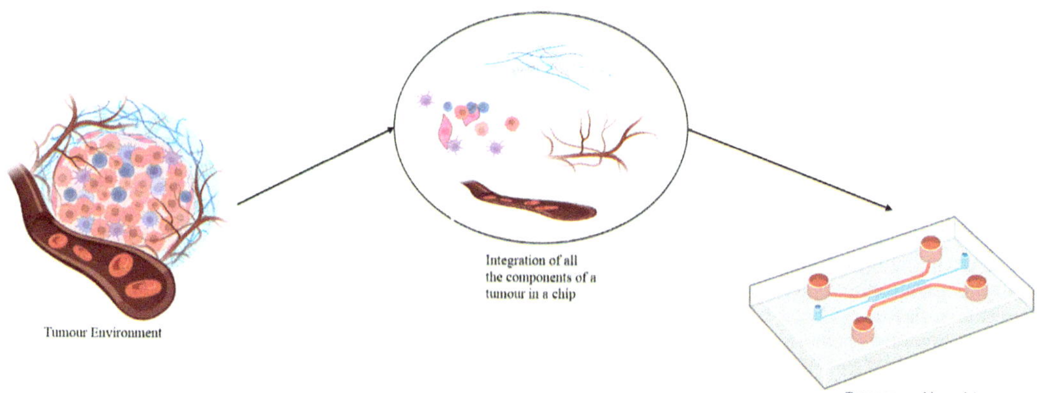

FIG. 9.4 Tumor-on-a-chip model. *From Biala, G., Kedzierska, E., Kruk-Slomka, M., Orzelska-Gorka, J., Hmaidan, S., Skrok, A., Kaminski, J., Havrankova, E., Nadaska, D., Malik, I., 2023. Research in the field of drug design and development. Pharmaceuticals 16, 1424-8247. doi:10.3390/ph16091283.*

allows researchers to mimic the mechanical properties of different tissue types, enabling a more realistic study of metastasis. This innovation not only improves the accuracy of preclinical drug testing but also reduces reliance on traditional animal models, accelerating the discovery of effective cancer treatments. This integration advances the development of efficient therapeutic options by enabling the investigation of dynamic processes under physiologically appropriate conditions, including drug reactions, invasion, and cell migration (Morwood et al., 2023; Saldin et al., 2017).

Another study introduced a liver-on-a-chip model to replicate the human TME and investigate drug efficacy and metabolism. The study examined statins, i.e., simvastatin, as a new anticancer drug for liver cancer. The findings revealed that statin metabolites derived from hepatocellular metabolism decreased the viability of human liver cancer cells; moreover, the cytotoxicity was selective to the cancerous cells, sparing primary hepatocytes and fibroblast cells (Chen et al., 2021). The model can be used as a platform for in vitro drug metabolism and toxicant studies, in addition to providing an accurate representation of the actual interior liver environment. In the near future, this LOC might be recognized as a platform for anticancer medication screening.

Recent advancements have enabled the incorporation of patient-derived cells into these microfluidic systems, allowing for personalized drug testing and therapy optimization. To create a more physiologically relevant tumor model, Shao and colleagues developed a novel organ-on-a-chip platform that combines an inverse opal scaffold with a microfluidic chip. They created the inverse opal scaffold by making a negative replica of an assembled microfluidic droplet template, which resulted in a 3D porous structure with uniform, interconnected pores. This design allows hepatoma cells to aggregate within the scaffold's pores, forming homogeneous tumor spheroids that closely resemble the architecture and microenvironment of in vivo tumors. They created liver tumor models with variable levels of drug resistance by injecting different proportions of drug-resistant and parental hepatoma cells onto the chip. This configuration allowed for the dynamic monitoring of tumor responses to chemotherapy over time, providing important information about how drug resistance develops over different tumor stages and treatment durations (Shao et al., 2023). By offering a more realistic setting for researching tumor dynamics and medication resistance, this organ-on-a-chip platform, in conclusion, represents a substantial breakthrough in cancer research and may eventually result in more effective cancer treatments.

References

Abbasi-Malati, Z., Khanicheragh, P., Narmi, M.T., Mardi, N., Khosrowshahi, N.D., Hiradfar, A., Rezabakhsh, A., Sadeghsoltani, F., Rashidi, S., Chegeni, S.A., Roozbahani, G., Rahbarghazi, R., 2024. Tumoroids, a valid preclinical screening platform for monitoring cancer angiogenesis. Stem Cell Res. Ther. 1757-6512. 15 (1). Available from: https://stemcellres.biomedcentral.com. doi:10.1186/s13287-024-03880-4.

Al-Akra, L., Bae, D.H., Leck, L.Y.W., Richardson, D.R., Jansson, P.J., 2019. The biochemical and molecular mechanisms involved in the role of tumor micro-environment stress in development of drug resistance. Biochim. Biophys. Acta Gen. Subj. 1872-8006. 1863 (9), 1390–1397. Available from: www.elsevier.com/locate/bbagen. doi:10.1016/j.bbagen.2019.06.007.

Alum, E.U., 2025. AI-driven biomarker discovery: enhancing precision in cancer diagnosis and prognosis. Uganda Discover Oncol. 2730-6011. 16 (1). Available from: https://www.springer.com/journal/12672. doi:10.1007/s12672-025-02064-7.

Biala, G., Kedzierska, E., Kruk-Slomka, M., Orzelska-Gorka, J., Hmaidan, S., Skrok, A., Kaminski, J., Havrankova, E., Nadaska, D., Malik, I., 2023. Research in the field of drug design and development. Pharmaceuticals. 1424-8247. 16 (9), 1283. doi:10.3390/ph16091283.

Blumer, T., Fofana, I., Matter, M.S., Wang, X., Montazeri, H., Calabrese, D., Coto-Llerena, M., Boldanova, T., Nuciforo, S., Kancherla, V., Tornillo, L., Piscuoglio, S., Wieland, S., Terracciano, L.M., Ng, C.K.Y., Heim, M.H., 2019. Hepatocellular carcinoma xenografts established from needle biopsies preserve the characteristics of the originating tumors. Hepatol. Commun. 2471-254X. 3 (7), 971–986. Available from: http://aasldpubs.onlinelibrary.wiley.com/hub/journal/10.1002/(ISSN)2471-254X/. doi:10.1002/hep4.1365.

Bruner, Derksen, 2018. Loss of E-cadherin-dependent cell–cell adhesion and the development and progression of cancer. Cold Spring Harb. Perspect. Biol. 10 (3).

Chatterjee, S., Bhat, V., Berdnikov, A., Liu, J., Zhang, G., Buchel, E., Safneck, J., Marshall, A.J., Murphy, L.C., Postovit, L.M., Raouf, A., 2019. Paracrine crosstalk between fibroblasts and ER^+ breast cancer cells creates an IL1β-Enriched niche that promotes tumor growth. iScience. 2589-0042. 19, 388–401. Available from: www.cell.com/iscience. doi:10.1016/j.isci.2019.07.034.

Chen, C., Wang, J., Pan, D., Wang, X., Xu, Y., Yan, J., Wang, L., Yang, X., Yang, M., Liu, G.P., 2023. Applications of multi-omics analysis in human diseases. MedComm. 2688-2663. 4 (4). Available from: onlinelibrary.wiley.com/journal/26882663. doi:10.1002/mco2.315.

Chen, P.-Y., Hsieh, M.-J., Liao, Y.-H., Lin, Y.-C., Hou, Y.-T., 2021. Liver-on-a-chip platform to study anticancer effect of statin and its metabolites. Biochem. Eng. J. 165, 107831. doi:10.1016/j.bej.2020.107831 .

Choi, P.J., Tubbs, R.S., Oskouian, R.J., n.d. The current trend of the translational research paradigm. Cureus 10 (3).

Creedon, H., Balderstone, L.A., Muir, M., Balla, J., Gomez-Cuadrado, L., Tracey, N., Loane, J., Klinowska, T., Muller, W.J., Brunton, V.G., 2016. Use of a genetically engineered mouse model as a preclinical tool for HER2 breast cancer. Dis. Model. Mech. 1754-8411. 9 (2), 131–140. Available from: http://dmm.biologists.org/content/dmm/9/2/131.full.pdf. doi:10.1242/dmm.023143.

De Maria Marchiano, R., Di Sante, G., Piro, G., Carbone, C., Tortora, G., Boldrini, L., Pietragalla, A., Daniele, G., Tredicine, M., Cesario, A., Valentini, V., Gallo, D., Babini, G., D'Oria, M., Scambia, G., 2021. Translational research in the era of precision medicine: Where we are and where we will go. J. Pers. Med. 11 (3), 216. doi:10.3390/jpm11030216.

Ding, Z., Wu, C.J., Chu, G.C., Xiao, Y., Ho, D., Zhang, J., Perry, S.R., Labrot, E.S., Wu, X., Lis, R., Hoshida, Y., Hiller, D., Hu, B., Jiang, S., Zheng, H., Stegh, A.H., Scott, K.L., Signoretti, S., Bardeesy, N., Wang, Y.A., Hill, D.E., Golub, T.R., Stampfer, M.J., Wong, W.H., Loda, M., Mucci, L., Chin, L., Depinho, R.A., 2011. SMAD4-dependent barrier constrains prostate cancer growth and metastatic progression. Nature. 1476-4687. 470 (7333), 269–276. doi:10.1038/nature09677.

Distler, M., Aust, D., Weitz, J., Pilarsky, C., Grützmann, R., 2014. Precursor lesions for sporadic pancreatic cancer: panIN, IPMN, and MCN. Biomed Res. Int. 2314-6133. 2014, 1–11. doi:10.1155/2014/474905.

Doublier, S., Belisario, D.C., Polimeni, M., Annaratone, L., Riganti, C., Allia, E., Ghigo, D., Bosia, A., Sapino, A., 2012. HIF-1 activation induces doxorubicin resistance in MCF7 3-D spheroids via P-glycoprotein expression: a potential model of the chemo-resistance of invasive micropapillary carcinoma of the breast. BMC Cancer. 1471-2407. 12. Available from: http://www.biomedcentral.com/1471-2407/12/4. doi:10.1186/1471-2407-12-4Italy.

Ebright, R.Y., Lee, S., Wittner, B.S., Niederhoffer, K.L., Nicholson, B.T., Bardia, A., Truesdell, S., Wiley, D.F., Wesley, B., Li, S., Mai, A., Aceto, N., Vincent-Jordan, N., Szabolcs, A., Chirn, B., Kreuzer, J., Comaills, V., Kalinich, M., Haas, W., Ting, D.T., Toner, M., Vasudevan, S., Haber, D.A., Maheswaran, S., Micalizzi, D.S., 2020. Deregulation of ribosomal protein expression and translation promotes breast cancer metastasis. Science. 1095-9203. 367 (6485), 1468–1473. Available from: https://science.sciencemag.org/content/sci/367/6485/1468.full.pdf. doi:10.1126/science.aay0939.

Echeverria, G.V., Ge, Z., Seth, S., Zhang, X., Jeter-Jones, S., Zhou, X., Cai, S., Tu, Y., McCoy, A., Peoples, M., Sun, Y., Qiu, H., Chang, Q., Bristow, C., Carugo, A., Shao, J., Ma, X., Harris, A., Mundi, P., Lau, R., Ramamoorthy, V., Wu, Y., Alvarez, M.J., Califano, A., Moulder, S.L., Symmans, W.F., Marszalek, J.R., Heffernan, T.P., Chang, J.T., Piwnica-Worms, H., 2019a. Resistance to neoadjuvant chemotherapy in triple-negative breast cancer mediated by a reversible drug-tolerant state. Sci. Transl. Med. 1946-6242. 11 (488). Available from: https://stm.sciencemag.org/content/scitransmed/11/488/eaav0936.full.pdf. doi:10.1126/scitranslmed.aav0936.

Echeverria, G.V., Ge, Z., Seth, S., Zhang, X., Jeter-Jones, S., Zhou, X., Cai, S., Tu, Y., McCoy, A., Peoples, M., Sun, Y., Qiu, H., Chang, Q., Bristow, C., Carugo, A., Shao, J., Ma, X., Harris, A., Mundi, P., Lau, R., Ramamoorthy, V., Wu, Y.,

Alvarez, M.J., Califano, A., Moulder, S.L., Symmans, W.F., Marszalek, J.R., Heffernan, T.P., Chang, J.T., Piwnica-Worms, H., 2019b. Resistance to neoadjuvant chemotherapy in triple-negative breast cancer mediated by a reversible drug-tolerant state. Sci. Transl. Med. 1946-6242. 11 (488). Available from: https://stm.sciencemag.org/content/scitransmed/11/488/eaav0936.full.pdf. doi:10.1126/scitranslmed.aav0936.

Elhamamsy, A.R., Metge, B.J., Alsheikh, H.A., Shevde, L.A., Samant, R.S., 2022. Ribosome biogenesis: a central player in cancer metastasis and therapeutic resistance. Cancer Res. 1538-7445. 82 (13), 2344–2353. Available from: https://aacrjournals.org/cancerres/article/82/13/2344/705027/Ribosome-Biogenesis-A-Central-Player-in-Cancer. doi:10.1158/0008-5472.CAN-21-4087.

Fang, Y., Eglen, R.M., 2017. Three-dimensional cell cultures in drug discovery and development. SLAS Discov. 2472-5560. 22 (5), 456–472. Available from: http://journals.sagepub.com/loi/jbx. doi:10.1177/1087057117696795.

Gruvberger-Saal, Bendahl, Saal, Laakso, Hegardt, Edén, et al., 2007. Estrogen receptor beta expression is associated with tamoxifen response in ERalpha-negative breast carcinoma. Clin. Cancer Res. 13 (7), 1987–1994.

Guan, X., Huang, S., 2022. Advances in the application of 3D tumor models in precision oncology and drug screening. Front. Bioeng. Biotechnol. 2296-4185. 10. Available from: http://journal.frontiersin.org/journal/bioengineering-and-biotechnology#archive. doi:10.3389/fbioe.2022.1021966.

Hingorani, S.R., Wang, L., Multani, A.S., Combs, C., Deramaudt, T.B., Hruban, R.H., Rustgi, A.K., Chang, S., Tuveson, D.A., 2005. Trp53R172H and KrasG12D cooperate to promote chromosomal instability and widely metastatic pancreatic ductal adenocarcinoma in mice. Cancer Cell. 1535-6108. 7 (5), 469–483. Available from: https://www.journals.elsevier.com/cancer-cell. doi:10.1016/j.ccr.2005.04.023.

Isaacs, J.T., Coffey, D.S., 1981. Adaptation versus selection as the mechanism responsible for the relapse of prostatic cancer to androgen ablation therapy as studied in the dunning R-3327-H adenocarcinoma. Cancer Res. 1538-7445. 41, 5070–5074.

Kersten, K., de Visser, K.E., van Miltenburg, M.H., Jonkers, J., 2017. Genetically engineered mouse models in oncology research and cancer medicine. EMBO Mol. Med. 1757-4684. 9 (2), 137–153. Available from: http://onlinelibrary.wiley.com/journal/10.1002/(ISSN)1757-4684. doi:10.15252/emmm.201606857.

Lampreht Tratar, U., Horvat, S., Cemazar, M., 2018. Transgenic mouse models in cancer research. Front. Oncol. 2234-943X. 8. doi:10.3389/fonc.2018.00268.

Leenaars, C.H.C., Kouwenaar, C., Stafleu, F.R., Bleich, A., Ritskes-Hoitinga, M., De Vries, R.B.M., Meijboom, F.L.B., 2019. Animal to human translation: a systematic scoping review of reported concordance rates. J. Transl. Med. 1479-5876. 17 (1). Available from: http://www.translational-medicine.com/home/. doi:10.1186/s12967-019-1976-2.

Liu, X., Fang, J., Huang, S., Wu, X., Xie, X., Wang, J., Liu, F., Zhang, M., Peng, Z., Hu, N., 2021. Tumor-on-a-chip: from bioinspired design to biomedical application. Microsyst. Nanoeng. 2055-7434. 7 (1). doi:10.1038/s41378-021-00277-8.

Liu, Y., Wu, W., Cai, C., Zhang, H., Shen, H., Han, Y., 2023. Patient-derived xenograft models in cancer therapy: technologies and applications. Signal Transduct. Target. Ther. 2059-3635. 8 (1). doi:10.1038/s41392-023-01419-2.

Lower, E.E., Glass, E., Blau, R., Harman, S., 2009. HER-2/neu expression in primary and metastatic breast cancer. Breast Cancer Res. Treat. 1573-7217. 113 (2), 301–306. doi:10.1007/s10549-008-9931-6.

Manduca, N., Maccafeo, E., De Maria, R., Sistigu, A., Musella, M., 2023. 3D cancer models: one step closer to in vitro human studies. Front. Immunol. 1664-3224. 14. Available from: https://www.frontiersin.org/journals/immunology#. doi:10.3389/fimmu.2023.1175503.

Mitrakas, A.G., Tsolou, A., Didaskalou, S., Karkaletsou, L., Efstathiou, C., Eftalitsidis, E., Marmanis, K., Koffa, M., 2023. Applications and advances of multicellular tumor spheroids: challenges in their development and analysis. Int. J. Mol. Sci. 1422-0067. 24 (8). Available from: http://www.mdpi.com/journal/ijms. doi:10.3390/ijms24086949.

Moody, S.E., Sarkisian, C.J., Hahn, K.T., Gunther, E.J., Pickup, S., Dugan, K.D., Innocent, N., Cardiff, R.D., Schnall, M.D., Chodosh, L.A., 2002. Conditional activation of Neu in the mammary epithelium of transgenic mice results in reversible pulmonary metastasis. Cancer Cell. 1535-6108. 2 (6), 451–461. Available from: https://www.journals.elsevier.com/cancer-cell. doi:10.1016/S1535-6108(02)00212-X.

Morwood, A.J., El-Karim, I.A., Clarke, S.A., Lundy, F.T., 2023. The role of extracellular matrix (ECM) adhesion motifs in functionalised hydrogels. Molecules. 1420-3049. 28 (12). Available from: http://www.mdpi.com/journal/molecules. doi:10.3390/molecules28124616.

Mosig, A.S., 2017. Organ-on-chip models: new opportunities for biomedical research. Future Sci. OA. 2056-5623. 3 (2). Available from: http://www.future-science.com/loi/fso?open=2015. doi:10.4155/fsoa-2016-0038.

Natu, J., Nagaraju, G.P., 2023. Gemcitabine effects on tumor microenvironment of pancreatic ductal adenocarcinoma: special focus on resistance mechanisms and metronomic therapies. Cancer Lett. 1872-7980. 573. Available from: www.elsevier.com/locate/canlet. doi:10.1016/j.canlet.2023.216382.

Okada, S., Vaeteewoottacharn, K., Kariya, R., 2018. Establishment of a patient-derived tumor xenograft model and application for precision cancer medicine. Chem. Pharm. Bull. 0009-2363. 66 (3), 225–230. doi:10.1248/cpb.c17-00789.

Park, C.K., Khalil, M., Pham, N.A., Wong, S., Ly, D., Sacher, A., Tsao, M.S., 2025. Humanized mouse models for immuno-oncology research: a review and implications in lung cancer research. JTO Clin. Res. Rep. 2666-3643. 6 (3). Available from: https://www.sciencedirect.com/science/journal/26663643. doi:10.1016/j.jtocrr.2024.100781.

Petrosino, J.F., 2018. The microbiome in precision medicine: the way forward. Genome Med. 1756-994X. 10 (1). Available from: http://www.genomemedicine.com/. doi:10.1186/s13073-018-0525-6.

Pinto, B., Henriques, A.C., Silva, P.M.A., Bousbaa, H., 2020. Three-dimensional spheroids as in vitro preclinical models for cancer research. Pharmaceutics. 1999-4923. 12 (12), 1–38. Available from: http://www.mdpi.com/journal/pharmaceutics. doi:10.3390/pharmaceutics12121186.

Pompili, L., Porru, M., Caruso, C., Biroccio, A., Leonetti, C., 2016. Patient-derived xenografts: a relevant preclinical model for drug development. J. Exp. Clin. Cancer Res. 1756-9966. 35 (1). Available from: http://www.jeccr.com/. doi:10.1186/s13046-016-0462-4.

R Marchiano, M.D., Di Sante, G., Piro, G., Carbone, C., Tortora, G., Boldrini, et al., 2021. Translational research in the era of precision medicine: where we are and where we will go. J. Pers. Med. 18 (3).

Researchgate, 2025. Translational research in drug discovery and development.

Rivlin, N., Brosh, R., Oren, M., Rotter, V., 2011. Mutations in the p53 tumor suppressor gene: important milestones at the various steps of tumorigenesis. Genes Cancer. 1947-6027. 2 (4), 466–474. doi:10.1177/1947601911408889.

Roy, M., Alix, C., Bouakaz, A., Serrière, S., Escoffre, J.M., 2023. Tumor spheroids as model to design acoustically mediated drug therapies: a review. Pharmaceutics. 1999-4923. 15 (3). Available from: http://www.mdpi.com/journal/pharmaceutics. doi:10.3390/pharmaceutics15030806.

Rubin, E.H., Gilliland, D.G., 2012. Drug development and clinical trials—the path to an approved cancer drug. Nat. Rev. Clin. Oncol. 9 (4), 215–222. doi:10.1038/nrclinonc.2012.22 .

Rubio, D.M.G., Schoenbaum, E.E., Lee, L.S., Schteingart, D.E., Marantz, P.R., Anderson, K.E., Platt, L.D., Baez, A., Esposito, K., 2010. Defining translational research: implications for training. Acad. Med. 1040-2446. 85 (3), 470–475. Available from: http://www.lww.com/product/?1040-2446. doi:10.1097/ACM.0b013e3181ccd618.

Sachs, N., de Ligt, J., Kopper, O., Gogola, E., Bounova, G., Weeber, F., Balgobind, A.V., Wind, K., Gracanin, A., Begthel, H., Korving, J., van Boxtel, R., Duarte, A.A., Lelieveld, D., van Hoeck, A., Ernst, R.F., Blokzijl, F., Nijman, I.J., Hoogstraat, M., van de Ven, M., Egan, D.A., Zinzalla, V., Moll, J., Boj, S.F., Voest, E.E., Wessels, L., van Diest, P.J., Rottenberg, S., Vries, R.G.J., Cuppen, E., Clevers, H., 2018. A living biobank of breast cancer organoids captures disease heterogeneity. Cell. 1097-4172. 172 (1-2), 373. Available from: https://www.sciencedirect.com/journal/cell. doi:10.1016/j.cell.2017.11.010.

Saldin, L.T., Cramer, M.C., Velankar, S.S., White, L.J., Badylak, S.F., 2017. Extracellular matrix hydrogels from decellularized tissues: structure and function. Acta Biomater. 1878-7568. 49, 1–15. Available from: http://www.journals.elsevier.com/acta-biomaterialia. doi:10.1016/j.actbio.2016.11.068.

Shao, C., Yu, Y., Lei, X., Cao, J., Zhao, Y., Ye, F., 2023. Organ-on-a-chip for dynamic tumor drug resistance investigation. Chem. Eng. J. 1385-8947. 460, 141739. doi:10.1016/j.cej.2023.141739.

Sharma, K., Dey, S., Karmakar, R., Rengan, A.K., 2024. A comprehensive review of 3D cancer models for drug screening and translational research. Cancer Innov. 2770-9183. 3 (1). Available from: onlinelibrary.wiley.com/journal/27709183. doi:10.1002/cai2.102.

Skardal, A., Devarasetty, M., Forsythe, S., Atala, A., Soker, S., 2016. A reductionist metastasis-on-a-chip platform for in vitro tumor progression modeling and drug screening. Biotechnol. Bioeng. 1097-0290. 113 (9), 2020–2032. Available from: www.interscience.wiley.com/jpages/0006-3592. doi:10.1002/bit.25950.

Sun, D., Gao, W., Hu, H., Zhou, S., 2022. Why 90% of clinical drug development fails and how to improve it? Acta Pharm. Sin. B. 2211-3843. 12 (7), 3049–3062. Available from: http://www.elsevier.com/wps/find/journaldescription.cws_home/725753/description#description. doi:10.1016/j.apsb.2022.02.002.

Swain, S.M., Shastry, M., Hamilton, E., 2023. Targeting HER2-positive breast cancer: advances and future directions. Nat. Rev. Drug Discov. 22 (2), 101–126. doi:10.1038/s41573-022-00579-0.

Swain, S.M., Shastry, M., Hamilton, E., 2023. Targeting HER2-positive breast cancer: advances and future directions. Nat. Rev. Drug Discov. 1474-1784. 22 (2), 101–126. Available from: https://www.nature.com/nrd/. doi:10.1038/s41573-022-00579-0.

Szymański, P., Markowicz, M., Mikiciuk-Olasik, E., 2012. Adaptation of high-throughput screening in drug discovery—toxicological screening tests. Int. J. Mol. Sci. 1422-0067. 13 (1), 427–452. doi:10.3390/ijms13010427.

Wege, A.K., 2018. Humanized mouse models for the preclinical assessment of cancer immunotherapy. BioDrugs. 1179-190X. 32 (3), 245–266. Available from: http://rd.springer.com/journal/40259. doi:10.1007/s40259-018-0275-4.

Wensink, G.E., Elias, S.G., Mullenders, J., Koopman, M., Boj, S.F., Kranenburg, O.W., Roodhart, J.M.L., 2021. Patient-derived organoids as a predictive biomarker for treatment response in cancer patients. NPJ Precis. Oncol. 5 (1), 30. doi:10.1038/s41698-021-00168-1 .

Wittenburg, L.A., Gustafson, D.L., 2011. Optimizing preclinical study design in oncology research. Chem. Biol. Interact. 0009-2797. 190 (2-3), 73–78. doi:10.1016/j.cbi.2011.01.029.

Wu, F., Zhou, Y., Li, L., Shen, X., Chen, G., Wang, X., Liang, X., Tan, M., Huang, Z., 2020. Computational approaches in preclinical studies on drug discovery and development. Front. Chem. 2296-2646. 8. Available from: http://journal.frontiersin.org/journal/chemistry. doi:10.3389/fchem.2020.00726.

Xie, J., Lin, Y., 2020. Patient-derived xenograft models for personalized medicine in colorectal cancer. Clin. Exp. Med. 1591-9528. 20 (2), 167–172. Available from: link.springer.de/link/service/journals/10238/index.htm. doi:10.1007/s10238-020-00609-4.

Yang, R., Yu, Y., 2023. Patient-derived organoids in translational oncology and drug screening. Cancer Lett. 0304-3835. 562, 216180. doi:10.1016/j.canlet.2023.216180.

Yang, S., Hu, H., Kung, H., Zou, R., Dai, Y., Hu, Y., Wang, T., Lv, T., Yu, J., Li, F., 2023. Organoids: the current status and biomedical applications. MedComm. 2688-2663. 4 (3). doi:10.1002/mco2.274.

Yoshida, G.J., 2020. Applications of patient-derived tumor xenograft models and tumor organoids. J. Hematol. Oncol. 1756-8722. 13 (1). Available from: http://www.jhoonline.org/. doi:10.1186/s13045-019-0829-z.

Yuan, Y., Liu, X., Cai, Y., Li, W., 2022. Lapatinib and lapatinib plus trastuzumab therapy versus trastuzumab therapy for HER2 positive breast cancer patients: an updated systematic review and meta-analysis. Syst. Rev. 2046-4053. 11 (1). Available from: https://systematicreviewsjournal.biomedcentral.com/. doi:10.1186/s13643-022-02134-9.

Zang, Y., Lee, J.J., 2014. Adaptive clinical trial designs in oncology. Chinese Clin. Oncol. 2304-3873. 3 (4). Available from: http://cco.amegroups.com/article/download/4210/5081. doi:10.3978/j.issn.2304-3865.2014.06.04.

Zhao, M., Mishra, L., Deng, C.X., 2018. The role of TGF-β/SMAD4 signaling in cancer. Int. J. Biol. Sci. 1449-2288. 14 (2), 111–123. Available from: http://www.ijbs.com/v14p0111.pdf. doi:10.7150/ijbs.23230.

Zhao, W., Li, X., Nian, W., Wang, J., Wang, X., Sun, L., Zhu, Y., Tong, Z., 2021. Ribosome proteins represented by RPL27A mark the development and metastasis of triple-negative breast cancer in mouse and human. Front. Cell Dev. Biol. 2296-634X. 9. Available from: https://www.frontiersin.org/journals/cell-and-developmental-biology#. doi:10.3389/fcell.2021.716730.

Zhao, Z., Chen, X., Dowbaj, A.M., Sljukic, A., Bratlie, K., Lin, L., 2022. Organoids. Nat. Rev. Methods Prim. 2 (1), 1–21.

Index

Note: Page numbers followed by "*f*" indicate figures, "*t*" indicate tables.

D

Printed and bound by CPI Group (UK) Ltd, Croydon, CR0 4YY
22/09/2026
02220476-0007